MW00993474
GT2RS

# LUXURY & SPEED

## WORLD'S GREATEST CARS

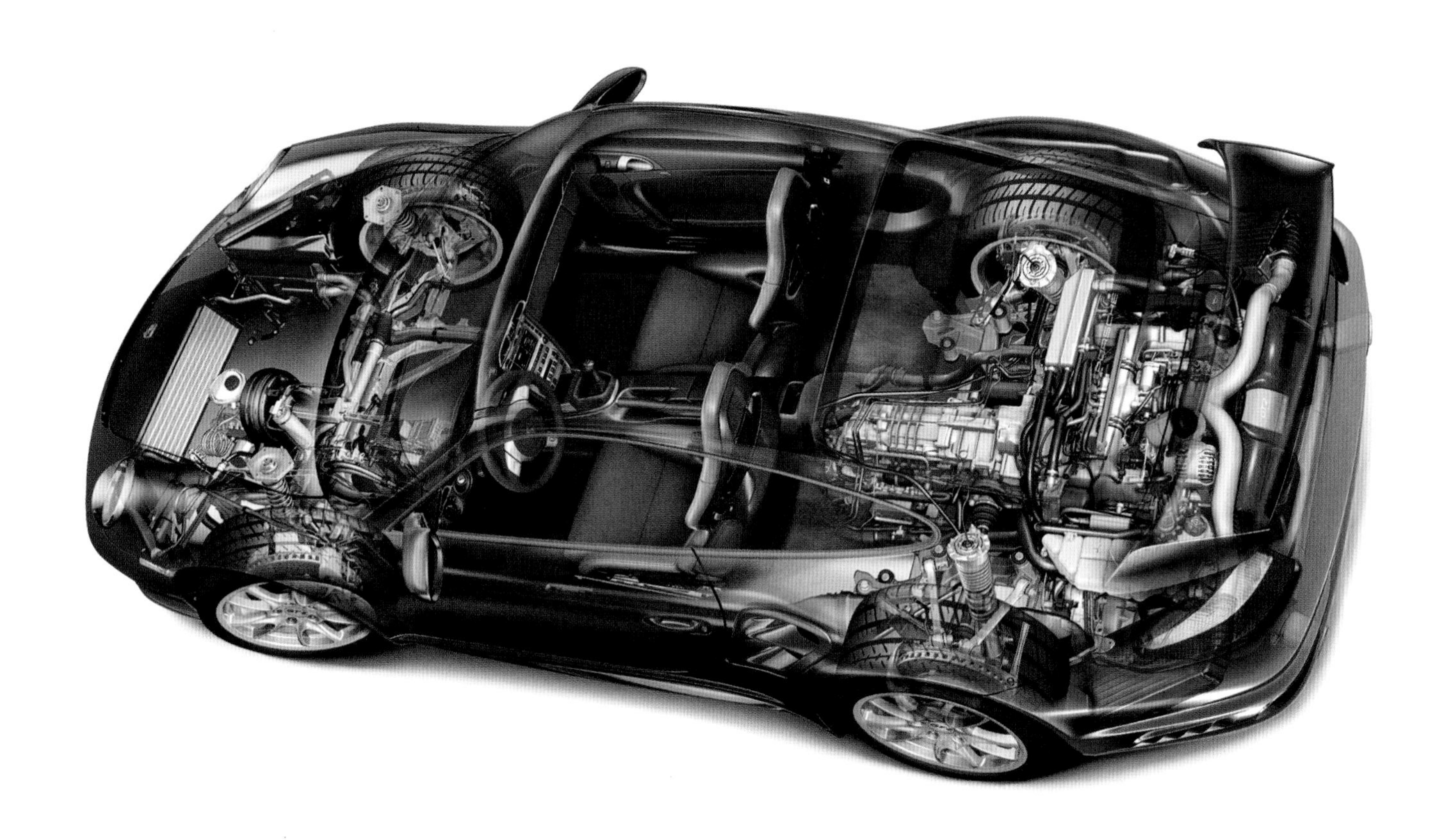

Publications International, Ltd.

Louis Weber, CEO
Publications International, Ltd.
8140 Lehigh Avenue
Morton Grove, IL 60053



ISBN: 978-1-64030-374-4

Manufactured in China.

8 7 6 5 4 3 2 1

# CREDITS

We would like to thank the following vehicle owners and photographers for supplying the images in this book.

**1907 Ford:** O: Larry O'Neal; P: Milton Kieft. **1913 Chevrolet:** O: Sloan Museum; P: Al Rogers. **1916 Scripps-Booth:** O: Kay Downing; P: W.C. Waymack. **1927 Mercedes-Benz:** O: Edward H. Wachs; P: Sam Griffith. **1928 Studebaker:** O: Brian McLaughlin; P: Vince Manocchi. **1929 Cadillac:** O: Thomas Tkacz; P: Leigh Dorrington. **1930 Cord:** O: Ed & Judy Schoenthaler; P: Doug Mitchel. **1930 Isotta Fraschini:** O: Paul Emple; P: Doug Mitchel. **1930 Lincoln:** O: Tim Sharon; P: Vince Manocchi. **1930 Packard:** O: Ed & Judy Schoenthaler; P: Doug Mitchel. **1930 Packard:** O: Fred R. Mauck; P: Vince Manocchi. **1931 Buick:** O: R.C. "Buzz" Pitzen; P: Vince Manocchi. **1931 Cadillac:** O: Steven Muccillo; P: Vince Manocchi. **1933 Delage:** O: Jacques Harguindeguy; P: Michael Scott. **1933 Duesenberg:** O: Ray Scherr; P: Vince Manocchi. **1935 Packard:** O: Walter G. & Shirley E. Sharpe; P: W.C. Waymack. **1936 Packard:** O: Ray Scherr; P: Vince Manocchi. **1937 Chrysler: O:** Robert Hepler; P: Vince Manocchi. **1937 Rolls-Royce:** O: Richard T. & Rexford Parker; P: Neil Nissing. **1938 Cadillac:** O: Richard Stanley; P: Vince Manocchi. **1938 Peugeot:** O: Scott Boses & Raymond Milo; P: Neil Nissing. **1939 Buick:** O: Erik Unthank; P: Vince Manocchi. **1940 Packard:** O: David E. Miller; P: Vince Manocchi. **1940 Packard:** O: D. David France; P: Vince Manocchi. **1941 Cadillac:** O: Ron Brooks, P: Doug Mitchel. **1941 Chrysler:** O: National Automobile Museum; P: Nicky Wright. **1941 Packard:** O: Roger Gibb; P: Vince Manocchi. **1941 Packard:** O: Neil Torrence; P: Vince Manocchi. **1947 Bentley:** O: Gary L. Wales; P: Vince Manocchi. **1947 Studebaker:** O: John Allen; P: Doug Mitchel. **1947 Triumph:** O: Gary & June Cooper; P: David Gooley. **1949 Cadillac:** O: Steve Plunkett; P: Al Rogers. **1949 Lincoln:** O: William A. Harper; P: Doug Mitchel. **1950 Chrysler:** O: Robert V. Huelsman; P: Al Rogers. **1951 Chrysler:** O: Art Astor; P: Vince Manocchi. **1951 Jowett:** O: David Burrows; P: Al Rogers. **1952 Healey:** O: Bryan Williams; P: Nicky Wright. **1952 Packard:** O: Ralph Marano; P: Don Heiny. **1953 Alfa Romeo:** O: Blackhawk Collection; P: Phil Toy. **1953 Cadillac:** O: Leonard Nagel; P: Vince Manocchi. **1953 Cadillac:** O: Petersen Automotive Museum; P: Vince Manocchi. **1953 Kaiser:** O: Richard Bailey; P: David Temple. **1954 Alfa Romeo:** O: Frank Gabrielli; P: Phil Toy. **1954 Chrysler:** O: Michael Pomerance; P: Don Heiny. **1955 Ford:** O: Ray Scherr; P: Vince Manocchi. **1955 Lincoln:** O: Tom Griffith; P: Thomas Glatch. **1956 DeSoto:** O: Bob Justice; P: Al Rogers. **1956 Maserati:** O: Johnathan Segal; P: David Gooley. **1956 Oldsmobile:** O: Jimmy Blackburn; P: David Temple. **1956 Packard:** O: Ralph & Kathy Tompkins; P: Doug Mitchel. **1956 Studebaker:** O: Mr. & Mrs. Ken Mooney; P: Nina Padgett. **1957 Buick:** O: Dennis Etcheverry; P: Phil Toy. **1957 Mercedes-Benz:** O: Bob Gunthrop; P: David Gooley. **1958 Ferrari:** O: Hilary Raab; P: Nicky Wright. **1959 Chrysler:** O: John Knab; P: Doug Mitchel. **1959 Jaguar:** O: Peter Rothenberg; P: David Gooley. **1960 Cadillac:** O: Bill Shipp; P: David Temple. **1961 Chevrolet:** O: Ken Lingenfelter; P: Al Rogers. **1962 Mercury:** O: Gary Richards; P: Vince Manocchi. **1963 Buick:** O: General Motors Company; P: Nicky Wright. **1963 Ferrari:** O: Frank & Leah Gabrielli; P: Phil Toy. **1963 Studebaker:** O: Gene Peiter; P: David Temple. **1966 Ford:** O: John W. Petras; P: Doug Mitchel. **1967 BMW:** O: Christian Weissmann; P: Vince Manocchi. **1968 Lincoln:** O: Donald W. Wesemann; P: Doug Mitchel. **1970 Jaguar:** O: Don Magargee; P: David Gooley. **1972 Oldsmobile:** O: Michael Calhoon; P: Tom Shaw. **1992-2000 Lexus:** O & P: Toyota Motor Corporation. **2001 Jaguar:** O & P: Jaguar Land Rover Limited. **2004-05 Cadillac:** O & P: General Motors Company. **2005 Ford:** O & P Ford Motor Company. **2008-10 Audi:** O & P: Audi AG. **2009-12 Cadillac:** O & P: General Motors Company. **2014-18 BMW:** O & P: BMW Group. **2018 Lexus:** O & P: Toyota Motor Corporation. **2018 McLaren:** O & P: McLaren Automotive. **2018 Porsche 911 GT2 RS:** O & P: Porsche **2019 Chevrolet:** O & P: General Motors Company.

Additional art from Shutterstock.com

# CONTENTS

# INTRODUCTION

*Luxury and Speed* tells the story of the automotive industry from the beginning of the twentieth century to today. Packard was the dominant American luxury car before World War II, but Cadillac was a tough challenger that gained the upper hand in the postwar years. Lincoln was also building luxurious, well-engineered cars. Duesenberg operated in a different sphere with speed, horsepower...and also price well in excess of the competition. The Depression killed Duesenberg and most of the remaining American independents. Overseas, Mercedes-Benz was selling supercharged sport touring cars, and Rolls-Royce specialized in silent luxury.

World War II accelerated technology. Cadillac engines and Hydra-Matic automatic transmissions were used in tanks, and the recently-developed Hydra-Matic returned from the war much improved. Italian performance cars made their presence known in the postwar years with Ferrari and Maserati becoming favorites of the jet set. Cadillac introduced a compact, overhead-valve V-8 that helped launch a horsepower race. Plus, the Chevrolet Corvette put that American V-8 muscle in a sports car.

Personal luxury cars came into prominence in the Sixties. These were luxurious-trimmed coupes that often had one of the brand's most powerful engines. With cheap gas and prosperity, engines grew in size and power. Drivers of big sedans and coupes sped comfortably along the highways with little regard for fuel economy. Those halcyon days ended with increasing emissions standards and two energy crises during the Seventies.

The Seventies and Eighties were dark years for performance, but a new golden age started to dawn in the Nineties as computer-controlled fuel injection and other advances led to a rebirth of performance. There was also an increased emphasis on interior fit and materials that meant greater luxury. Today, there is a new golden age of performance, with comfortable sedans able to effortlessly post performance figures that would shame some of the Sixties' hottest muscle cars.

Each car in this book has its own part in the automotive history. Read and see the story unfold.

1907
FORD
MODEL K
ROADSTER

From his start in the auto business, Henry Ford wanted to build affordable cars. His chief financial backer, a coal merchant named Alexander Malcomson, wanted to add luxury cars to the lineup. Even though the simple, low-cost Model T would later put America on wheels and make Henry rich, Malcomson's plan made sense in 1905. Packard and Cadillac both started out building small, affordable cars but added high-profit luxury four-cylinder cars and prospered.

Ford's first stab at the luxury market was the four-cylinder Model B. It didn't sell. His next attempt was the six-cylinder Model K. It didn't sell well either, although Ford's cheaper models were a big success.

On paper at least, the Model K had a lot going for it. Introduced in late 1905 as a 1906 model, it was one of the first sixes in America. It developed 40 bhp from 405 cid, and the roadster was capable of 60 mph, "A mile a min-

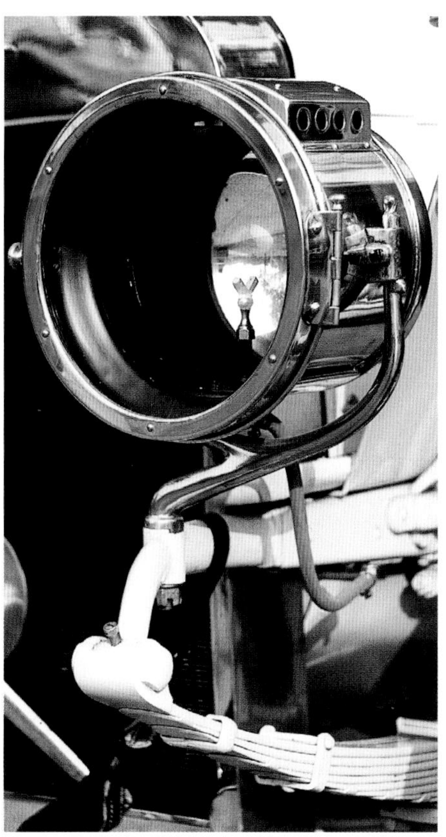

ute" was a heady claim before World War I. Priced $2500-$2800 for both roadster and touring car, it undercut the price of a four-cylinder Packard by more than $1000. Its weakness was the two-speed planetary transmission. Although durable in millions of later Model Ts , the transmission wasn't up to the task of handling the power of a big six.

Conventional history says that Henry hated the Model K and six-cylinder engines. (It was 1941 before Ford would build another six.) However, even after Malcomson, its patron, was forced out of the company as K production began, the car was sold into 1908. If Henry "hated" the car, why did he keep in production so long? Also, why design a six when fours were the norm? Perhaps Henry's dislike of the Model K and six-cylinder engines developed after the financial failure of the K.

Chevrolet

# 1913 CHEVROLET
## TYPE C TOURER

The first Chevrolet doesn't seem much like a Chevy. It wasn't reasonably priced or modestly sized. It didn't have an overhead-valve engine. It didn't even wear a bowtie badge.

The Chevrolet Type C, also known as the Classic Six, was the product of two conflicting personalities with some shared history. William Durant was a dealmaker. Fresh from his first ouster from General Motors in 1910, he was ready to establish a new automotive empire. Louis Chevrolet was what we'd now call a "car guy." A winning driver for Buick's racing team when Durant owned the company, he now wanted to design a car of his own. Durant wanted to build a light, inexpensive car; Chevrolet dreamed of a grand machine to proudly bear his name. In 1911, with Durant's backing, Chevrolet started working on a prototype in a Detroit factory. What developed was bigger and more expensive than Durant's vision.

When the Classic Six touring car entered production in late 1912 as a 1913 model, it was powered by a big 299-cid six and rode on a long 120-inch wheelbase. The price was a hefty $2150, and it probably cost more than that to build. However, Louis Chevrolet was justly proud of his car.

The engine was a T-head, which was a side-valve engine with intake and exhaust valves on opposite sides of the cylinder. This arrangement was favored for performance cars of the time. The crankshaft was well balanced and had counterweights so that it ran smoothly. The engine developed 40 bhp, enough to push the 3350-pound car up to 65 mph. The Classic Six was well equipped with electric lighting and a compressed-air starter that eliminated the need for manual cranking.

The speedometer was lighted to ease night driving, and there was a fuel gauge on the dash. Most cars at the time either didn't have a gas gauge or had one on the tank, not in easy view of the driver. The Classic Six was built to last and perhaps that contributed to its substantial weight. Satisfied with his namesake car, Louis Chevrolet left on a trip to his native Switzerland.

While Chevrolet was meticulously developing his ideal car, Durant had been moving much faster to bring popularly priced cars to market. A

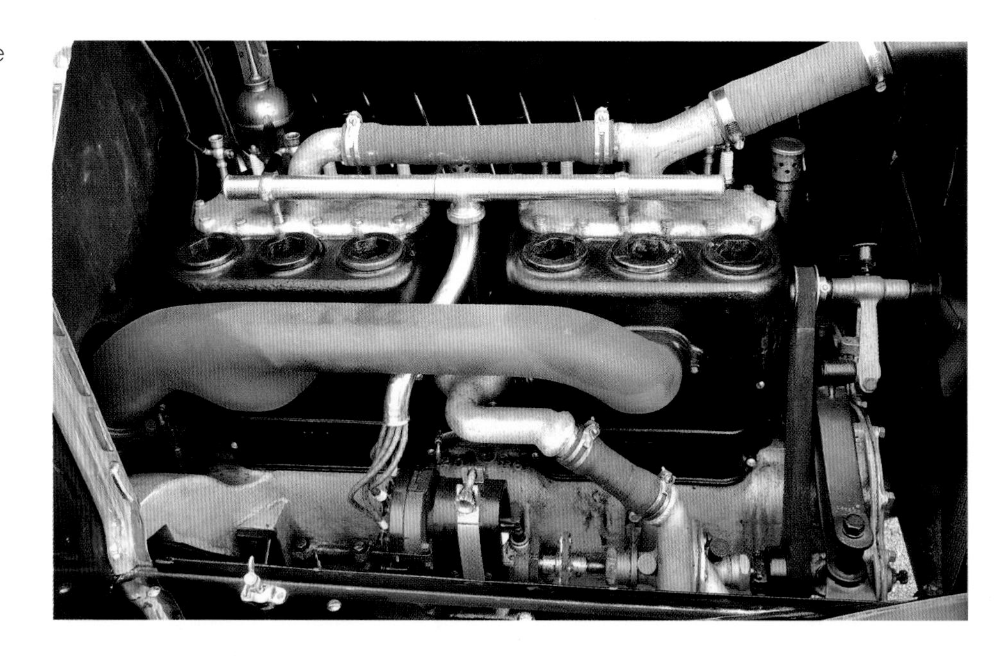

Little Four (named for William Little, a former Buick general manager) was on sale in 1912, and a Little Six made its debut at the same time as the Classic Six. When Chevrolet returned from his trip, he discovered that production of the slow-selling Classic Six had been moved from Detroit to the Little plant in Flint, Michigan. What's more, the Little Six became the Chevrolet Special Little Six, and four-cylinder Chevys were on the way. Chevrolet wasn't happy and legend has it that he finally stormed out

of the company when Durant told him to start smoking cigars instead of less-dignified cigarettes.

Classic Six production ended in 1914. Meanwhile, low-cost Chevrolets made Durant rich enough to buy his way back into the leadership of GM in 1915, and he established the mass-market brand that we know today.

This Classic Six is the oldest running Chevrolet. It bears body number 323, and was built after Chevrolet's move to the factory in Flint.

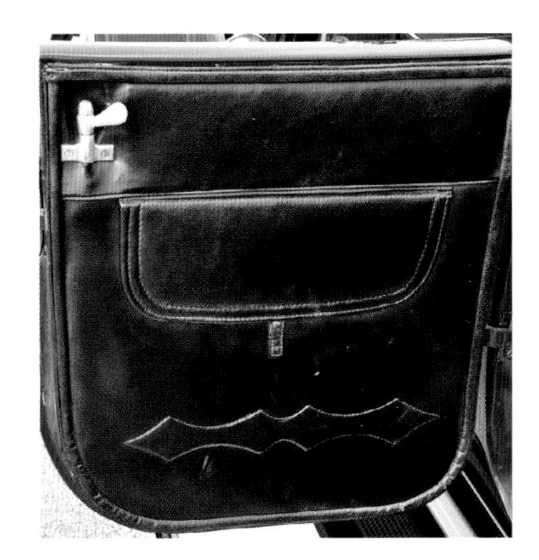

# 1916 SCRIPPS-BOOTH
## MODEL C ROADSTER

James Scripps Booth drew cars in his notebooks when he should have been paying attention in class. Booth might have failed the 10th grade three times before quitting school, but he was one of the few able to turn his automotive dreams into reality. He was not only blessed with access to financing (his grandfather founded the *Detroit News* and the Scripps-Howard newspaper chain) but he had a good grasp of engineering and marketing.

With financing from two uncles, James Booth set up a company in late 1914. William B. Stout, an engineer, was hired to make Booth's ideas and drawings into an automobile. Stout would later gain fame for his work on the Ford Tri-Motor passenger plane and the streamlined Stout Scarab car of the Thirties. Some consider the Scarab the first minivan.

The Scripps-Booth Model C had several unique features. Scripps was one of the first to mount the horn button on the steering wheel hub. Long before Hudson built its "Step-down" models, the Scripps-Booth's floor was lower than chassis rails to allow for a lower center of gravity and seating. The driver's seat was staggered ahead of the passenger seat for more elbow room in the narrow car. A jump seat folded down in front of

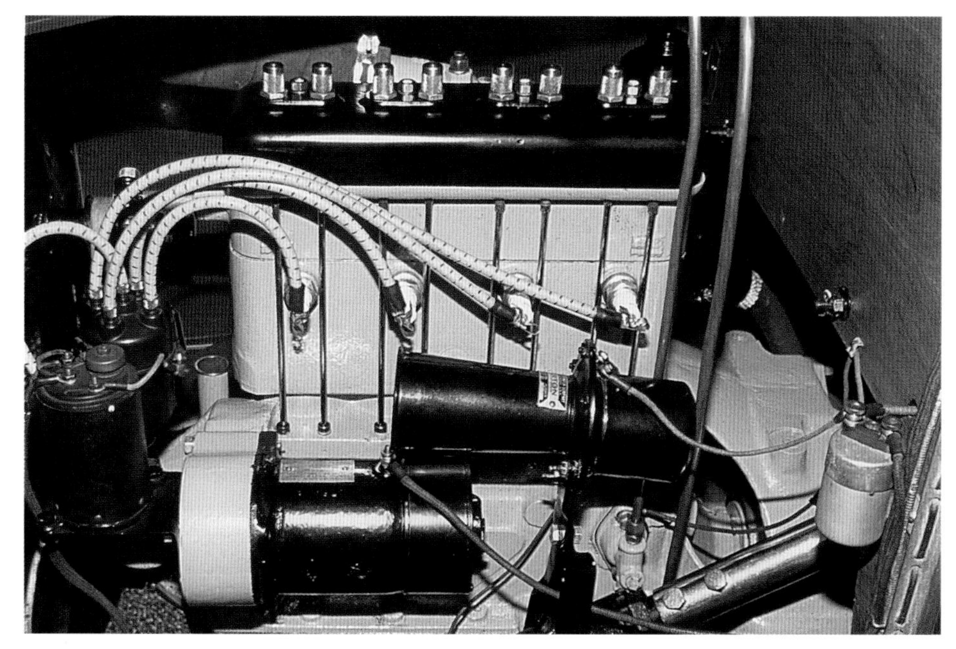

SCRIPPS
BOOTH
SB
DETROIT

the passenger seat for a third passenger. Power was provided by a 20-bhp ohv four-cylinder engine that later proved troublesome. Chevrolet engines were later substituted.

The Scripps-Booth car was advertised as a "luxurious light car." Although small and, at $775, not that expensive, Scripps-Booths were attractively styled and well-finished. They were also well-equipped with standard electric starting and high-quality Houk wire wheels. Marketing was aimed at the affluent. Scripps were often sold as second cars to owners of big luxury cars. The small, agile Scripps was just the thing for errands or joyrides in the country. The rich and famous were seen in Scripps-Booths

not only in America but also overseas. A third of production was exported, and Scripps were found in several royal garages. The car was fairly successful, and 6000 Model Cs were sold during 1915-16. More could have been sold if production had been able to keep up with demand. Scripps-Booth was at the mercy of its unreliable suppliers.

James Booth left the company in 1916, and it was soon in financial trouble. Reliability and production problems had caught up with the firm. A fast, V-8-powered Model D failed to revive sales. William Durant gained control of Scripps-Booth in 1917. By 1918, Scripps-Booths were built with General Motors components and were no longer

light cars. GM killed the make in 1922. Eventually, James Booth became a successful painter. He still found time to develop two auto prototypes, but they never made it to production.

# 1927 MERCEDES-BENZ
## S ROADSTER

Two significant events happened at Daimler, the maker of Mercedes cars, in 1923. That year it built the first production supercharged engines, and it added Ferdinand Porsche to the payroll.

Porsche applied supercharging to a new ohc six-cylinder engine. Mercedes supercharging was unconventional. Instead of pulling the fuel mixture from the carburetor, it blew through the carburetor. Also, the supercharger only engaged when the accelerator was floored, and the sound was loud enough to wake the dead. In fact,

36
73
MICHIGAN MIGLIA
1000 MILE CONTINENTAL RALLYE
310 AV
DEC 94

Mercedes advised against using the supercharger for more than 20 seconds.

The supercharged six was first used in a large luxury-car chassis, then a sportier K version with a shortened wheelbase was introduced in 1926—the year that the Daimler and Benz firms merged to form Daimler-Benz and sell Mercedes-Benz cars. The model name was derived from *kurz*, the German word for short. However, the K chassis was too high to be truly sporty so a low-chassis S (Sport) was introduced in 1927. The S retained the K's 133-inch wheelbase, but the engine was mounted farther back in the chassis for improved weight distribution.

The 6.8-liter six developed 120 bhp under natural aspiration or 180 bhp with the supercharger engaged. Top speed was more than 100 mph.

An S won the inaugural race at Germany's famous Nürburgring course in the car's competition debut, and it spawned even faster cars. The SS (Super Sport) had a bigger 7.1-liter engine. In 1928, the SSK put that larger engine in a shorter car. Also that year, Porsche quarreled with M-B management and left the company.

The final version was the SSKL, the "L" representing *leicht*, German for light. The SSKL had holes drilled in the chassis and used other weight-reduction measures to remove 250 pounds—although at around 3500 pounds it wasn't exactly svelte for a sports touring car. The SSKL was primarily for racing, and with its engine tuned to produce 300 bhp, it was capable of 146 mph. Production totals were 146 S, 111 SS, 33 SSK, and only a few units of the SSKL.

Before World War II, it wasn't unusual for owners of expensive cars to have a chassis rebodied. This car has skirted rear fenders that suggest the body might have been updated in the mid Thirties.

36

# 1928 STUDEBAKER
## PRESIDENT FOUR-DOOR SEDAN

Staid Studebaker loosened up with the Jazz Age. First, the uninspired model names Big Six, Special Six, and Standard Six were replaced by President, Commander, and the dubiously dubbed Dictator during 1927.

Then, Studebaker's ambitious head, Albert Erskine, wanted to build "the finest eight-cylinder car in the world regardless of cost." With the President Eight, he created an upper-medium-priced car that could challenge the luxury cars for power, comfort, and style.

The new eight was ready for 1928. Designed under the direction of Barney Roos, the 313-cid L-head engine produced 100 bhp, but was enlarged midyear to 337 cid with horsepower upped to 109.

Studebaker's Big Six held several speed and endurance records, and the company wanted to prove the new eight was even better. Two President sedans (similar to the car featured) and two roadsters were driven to the Atlantic City Speedway, a board track in New Jersey. The strictly stock cars circled the track for 20 days. At the end of the run, each car had covered 30,000 miles at average speeds of 63 mph for the slowest sedan and 68 mph for the fastest roadster. The American Automobile Association supervised the test and confirmed the only engine maintenance was a change of spark plugs and replacement of a fan belt.

Combined with previous tests, Studebaker held 114 stock-car records—some of which stood for 35 years.

Studebaker mounted its new engine in a long 131-inch-wheelbase FA chassis and a shorter 121-inch FB version. With attractive styling and luxurious interiors, the President was a success: More than 16,500 cars were sold in '28.

Today, the 1929-33 President is the only Studebaker recognized by the Classic Car Club of America. This 1928 five-passenger sedan is similar to the 1929 model, but inexplicably is not accorded Classic status.

## 1929 CADILLAC
### SERIES 341B FOUR-DOOR TOWN SEDAN

Cadillac was due for a "quiet" year in 1929, considering the previous season had seen the arrival of all-new Series 341 cars. They came with attractive styling by Harley Earl, fresh off his success with the 1927 LaSalle; a new 90-bhp 341-cid version of Caddy's established L-head V-8 engine; and a chassis with wheelbase stretched to 140 inches, a switch to torque-tube drive, and the adoption of 32-inch-diameter tires.

Those core features were indeed carried into 1929, but Cadillac was still able to make some noise in the automotive field—ironically by quieting things down. The marque introduced the "Synchro-Mesh" transmission, which made it easier to shift into second or

Cadillac

A·7809
CONN.

third gear without a gratingly audible clash of metal. Other improvements for '29 included internal-expanding brakes at all four wheels (external-contracting bands had been used previously at the rear), double-acting Delco shock absorbers, safety glass, and an adjustable front seat on most models. Fender-top parking lights replaced cowl lamps.

The 1929 Series 341B came in 11 "standard" models with Fisher bodies, plus another dozen "Fleetwood Custom" styles. Among the former was the "Town Sedan" featured here, a close-coupled five-passenger four-door model with a shortened body that made room at the back for a large detachable trunk.

The Cadillac is stunning in its original Pewter Pot, Blue Gray, and Black finish. But it is perhaps the interior details that are most alluring, including the ornate window regulators and door handles, rearview mirror, and dome light; ash receivers set into the door panels; robe rail; and assist straps. The crowning detail, however, is the original broadlace trim that was carefully preserved and returned to the door panels and seatback. When new, the Town Sedan was priced at $3495 to start. Options could add anywhere from $12.50 for the "herald" radiator ornament to $250 for dual sidemount spares and six Buffalo wire wheels. Other accessories include rear and side window shades, aftermarket vases, and a sealskin lap robe.

# 1930 CORD L-29 BROOKS STEVENS SPEEDSTER

The name Brooks Stevens is indelible in the annals of car design. Schooled in architecture in the Thirties, Stevens returned to his native Milwaukee, Wisconsin, to establish an industrial design firm. Like many others in the then-budding field, his clients included automakers, and over the years his company was responsible for the styling of numerous vehicles, such as the Willys Jeepster, the Jeep Wagoneer, and the 1962-64 GT Hawk update of Studebaker's legendary "Loewy coupe" design. Then, of course, there was the Excalibur, Stevens's mid-Sixties revival of Twenties classics themes.

Brooks Stevens's first car design wasn't for a client, however; it was for himself. The result was the Cord "speedster" seen on these pages. It was a particular favorite of the designer, who still had it at the time of his death in 1995.

Stevens became enthralled with E. L. Cord's new front-wheel-drive automobile when, as a teenager, he attended a wedding at which the newlyweds were given an L-29 as a present. Stevens's father, a successful engineer, helped him buy a two-year-old cabriolet as a reward for hard work.

In the mid-Thirties, about the time his studies at Cornell University in New York were completed, Brooks Stevens got the urge to customize his Cord convertible. The goal was to turn it into a speedster along the lines of those being built on Auburn, Duesenberg, and other prestige-car chassis.

The cowl was lowered slightly and the body narrowed. The overall shape of the cabriolet deck was retained, but the rumble seat was removed and the deck smoothed over, then capped with a trailing fin. A raked, vee'd windshield replaced the upright one-piece unit used on standard L-29s. The front fenders were skirted and the usual sidemount spares discarded, their fender wells filled in. Standard hood louvers were replaced by Duesenberg-like mesh screens that aided engine cooling when Stevens entered the car in hillclimbs. The exterior was capped by a Stevens-designed radiator cap, narrow Woodlite headlights, a rear bumper designed to protect the fin, and a sweeping two-tone paint scheme.

Inside, the engine-turned dash used an Auburn instrument panel—a neat layout that was a huge improvement on the L-29's gimmicky, hard-to-read design. The seatback was hinged so that when folded, there was access to the luggage compartment under the enclosed deck. Limousine Body, the Cord-owned coachbuilding shop, performed the modifications.

Stevens retained the stock 299-cid Lycoming straight-eight engine and three-speed transmission, the latter with its dash-mounted shift lever and distinctive "backward" shift pattern.

However, he had a twin-carburetor intake manifold and a special exhaust manifold made, and racier final-drive

gearing was installed. (A factory rebuilt engine was installed prior to the car's competition years.)

In later years, Stevens took his cherished Cord to shows and vintage-car touring events.

# 1930 ISOTTA FRASCHINI

## 8A FLYING STAR ROADSTER

Kings, tycoons, popes, and movie stars rode in Isotta Fraschinis. Clara Bow had an Isotta, and Rudolph Valentino owned two. The 1950 film *Sunset Boulevard* appropriately equipped fictional silent-film star Norma Desmond with an Isotta Fraschini landaulet.

The Italian automaker was best known for its big eights of the Twenties and Thirties, but Isotta Fraschini also built a broad range of sporting machines before World War I and was active in racing. In 1910, it was among the first to put brakes on all four wheels; most other makes didn't offer four-wheel brakes until the mid Twenties.

After the Great War, Isotta changed course and offered a single luxury model powered by the first production straight eight. Luxury-car buyers wanted smooth, flexible multicylinder engines, but the early V-8 engines had vibration issues. Until V-8 technology advanced, straight eights were smoother and dominated the luxury-car market between the wars. Isotta's 5.9-liter eight with overhead valves developed 80 bhp in the *Tipo* 8. In 1925, capacity was increased to 7.3 liters for the *Tipo* 8A and horsepower rose to 110-120. The power ratings are deceptive because there was also a tremendous amount of torque. The big Isotta could creep along at four mph in top gear of its three-speed transmission, and first gear was seldom needed as the car could easily launch in second. Top speed was more than 85 mph, and acceleration was strong. More important to Isotta's clientele was the eight's smoothness and reliability.

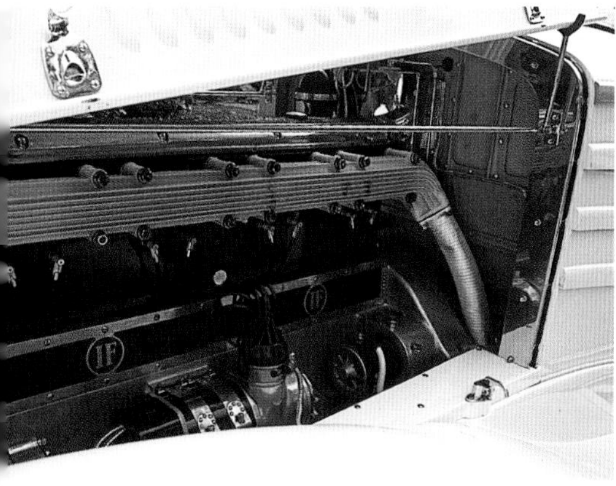

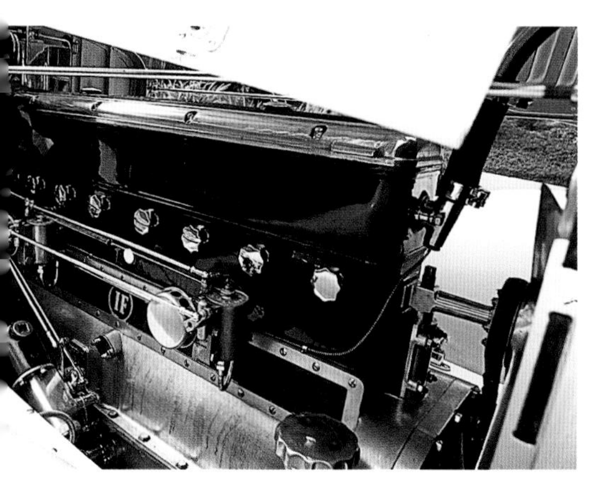

Isotta Fraschini competed with Rolls-Royce and Hispano-Suiza in the super-luxury market. Isottas were as meticulously crafted and as well-finished as their lofty competition. But it was the long hood and Italian coachwork on a standard 145-inch wheelbase that really set Isotta Fraschinis apart from other luxury cars. With the American economy booming in the Twenties, a third of the 1480 Isotta straight eights were sold here. An 8A chassis cost $6500 and custom bodies started at $6000, although a complete car could exceed $20,000—this when Lincoln's 1929 offerings topped out at $7400.

In 1928, the company began to concentrate on aircraft engine production. Then the 1929 stock market crash killed the lucrative American market. Dwindling car production ended in 1935. Prototypes for a rear-engined V-8 car were shown after World War II, but it was not put into production.

In 1931, Carrozzeria Touring built a striking Flying Star roadster on a short 135-inch 8A chassis. The car seen here originally had a coupe body, but that was replaced by an Australian reproduction of the Flying Star.

What of the original? Rumor has it that when last seen, it was being driven by Italian dictator Benito Mussolini and his mistress.

CALIFORNIA
HISTORICAL
VEHICLE
097E

# 1930 LINCOLN
## MODEL L JUDKINS BERLINE

A fast car for the Twenties, the Model L Lincoln was popular with both gangsters and policemen. Introduced in 1921, the Lincoln V-8 developed 90 bhp. The engine grew to 384.8 cubic inches in 1928, although horsepower officially remained the same. Acceleration was strong, and top speed for all but the heaviest bodystyles was around 90 mph.

Nineteen thirty would be the last year for the Model L. The Model K of '31 had a new longer and lower chassis but retained an improved version of the faithful Lincoln V-8.

Henry Leland, the father of Cadillac, also founded Lincoln. Leland insisted on the highest standards of engineering and construction. Unfortunately he didn't put much effort into the appearance of the new Lincoln. Dull styling and a recession led to receivership. Henry Ford bought Lincoln in 1922. Leland left in a huff that same year and Ford's son, Edsel, took over Lincoln management.

Edsel was good for Lincoln. Leland had created an outstanding chassis, but Ford had the taste needed to make the car look good.

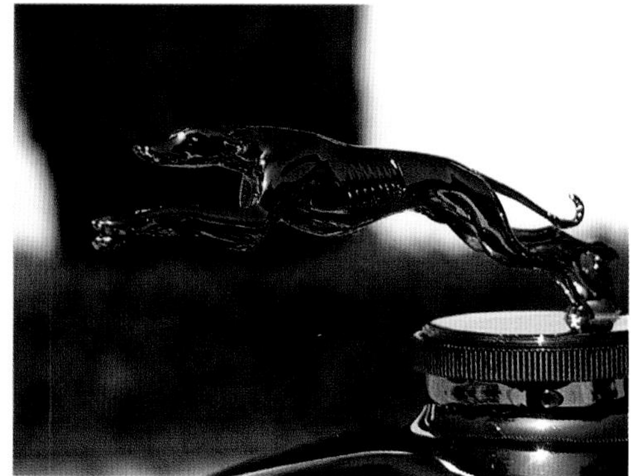

In 1925, Gorham silversmiths designed a chrome greyhound mascot for the radiator cap. It was molded using the "lost wax" process, a method that gave consistently fine detail. Rolls-Royce also used lost wax for its Spirit of Ecstasy.

Edsel enhanced Lincoln's image by ordering custom bodies from the leading coachbuilders in lots of 10 to more than 100. This provided distinctive coachwork at a more reasonable price than one-of-a-kind custom bodies. The Series 172 berline by Judkins shown here was one of the custom offerings. The word "berline" is derived from the German city Berlin. It was another term for a sedan—often a seven-passenger style with a divider between the front and rear compartments.

While a standard sedan cost $4500, berline prices started at $5600. Judkins built four versions of the berline. This 172-A with two side windows was one of 42 built in 1930. It features a distinctively angled windshield invented by Brewster coachbuilders. The configuration was thought to reduce glare and improve visibility in rain. It found some popularity in the Twenties but had almost disappeared by 1930.

This car was found hidden in a Chicago warehouse with an apparent bullet hole in the divider window, suggesting it had an interesting past.

## 1930
# PACKARD
734 SPEEDSTER
RUNABOUT

Unlike some of its peers in the top rank of American motordom, the Packard Motor Car Company wasn't loath to displaying flashes of flamboyance in the design or engineering of its products. A case in point is the 734 Speedster series of 1930, particularly the boattailed roadster version seen on these pages.

To create the 734 line, Packard started with the 134.5-inch-wheelbase chassis of its intermediate series, the 733. To it was added a highly modified version of the 384.8-cid L-head straight-eight that powered the larger 740- and 745-series Packards. In those cars, the engine generated 106 bhp. For use in the 734s, though, the power plant was given new separate intake and exhaust manifolds, enlarged exhaust valves, and bigger ports. With a 6.0:1-compression head, the engine made 145 bhp. (Even with the tamer optional 4.85:1 head the Speedster was still good for 125 horses.)

The transmission was, for the first time, a four-speed, the added cog being an ultra-low first gear for smoother starts. Behind the 19-inch wire wheels sat massive 16-inch mechanically actuated forged drum brakes borrowed from the 740/745, but with cooling fins specific to the 734 Speedsters.

Packard turned to its own custom body department for the lowered and narrowed coachwork of the five cataloged Speedsters: roadster with rumbleseat, Victoria convertible, sedan, phaeton, and two-passenger boattail roadster, the last of which Packard called the Runabout.

The Speedsters proved to be a fleeting extravagance in Depression-ravaged America. Base prices ranged from $5200 to $6000. The Runabout started at $5210. Announced in late January 1930, the 734s were produced only until the end of April. Perhaps 120 of all types were built.

PACKARDS
CALIFORNIA
HISTORICAL
VEHICLE
5833

# 1930 PACKARD
## DELUXE EIGHT ROADSTER

Flush with success from a prosperous fiscal-year 1929, the managers, stockholders, and customers of the Packard Motor Car Company had to feel confident as the new Seventh Series cars began to come out that autumn. After all, the venerable company had just recorded net sales of $107.5 million and a profit of nearly $26 million—both record highs for Packard—and the new models were in the wings with a touch of added beauty for 1930.

No one counted on "The Crash," however, the great Wall Street stock panic that October, which signaled the start of the Great Depression and the end of the palmy days at Packard. Its finances would never be as vital again, and the nature of its product offerings at the end of the Thirties would be quite different.

But that was all still to come when the Seventh Series was announced. At the time, Packards consisted of the Series 726 and Series 733 Standard Eights on respective 127.5-and 134.5-inch wheelbases, the Series 740 Custom Eight on a 140.5-inch spread, and the Series 745 DeLuxe Eight on a 145.5-inch chassis. (Racy and extremely rare Series 734 Speedster Eights on a 134-inch wheelbase were added in January.) There was some model shuffling, especially within the Standard Eight ranks, but the introductory 1930 lineup was essentially the same as 1929's Sixth Series.

PACKARDS
CALIFORNIA
HISTORICAL
VEHICLE
5833

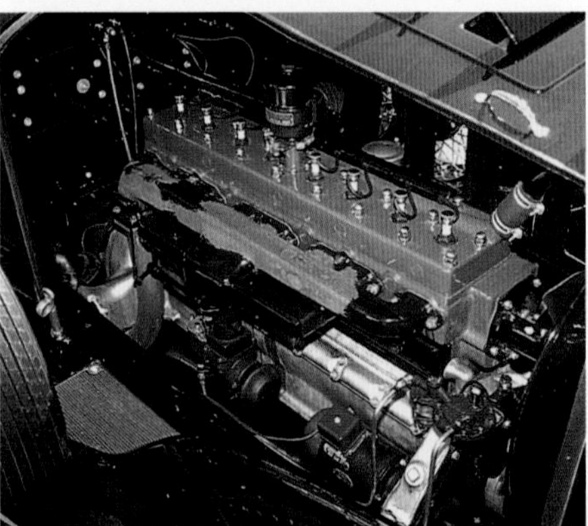

The $4585 roadster was one of 11 Series 745 factory bodies shared with the 733 and 740. (Design consultant Raymond Dietrich made the 1930 bodies look a bit lower.) The 740 and 745 made use of the same 384.8-cid L-head inline eight-cylinder engine of 106 bhp. A newly added "low-low" first gear turned the Packard transmission into a four-speed.

The difference in length between the 740 and 745 showed in the latter's five-inch-longer hood. The 745s were also distinguished by more sweeping front fenders, standard dual spare tires, and a "spur" that stuck out of the beltline molding just behind the radiator. Fender-mounted parking lights were a new touch for 740s and 745s.

This Series 745 roadster sports wire wheels—a factory option to the standard disc wheels—and an accessory trunk. It was exceedingly rare, even when new: Just 1789 of all Series 745 cars were built.

# 1931
# BUICK 95
## SEVEN-PASSENGER PHAETON

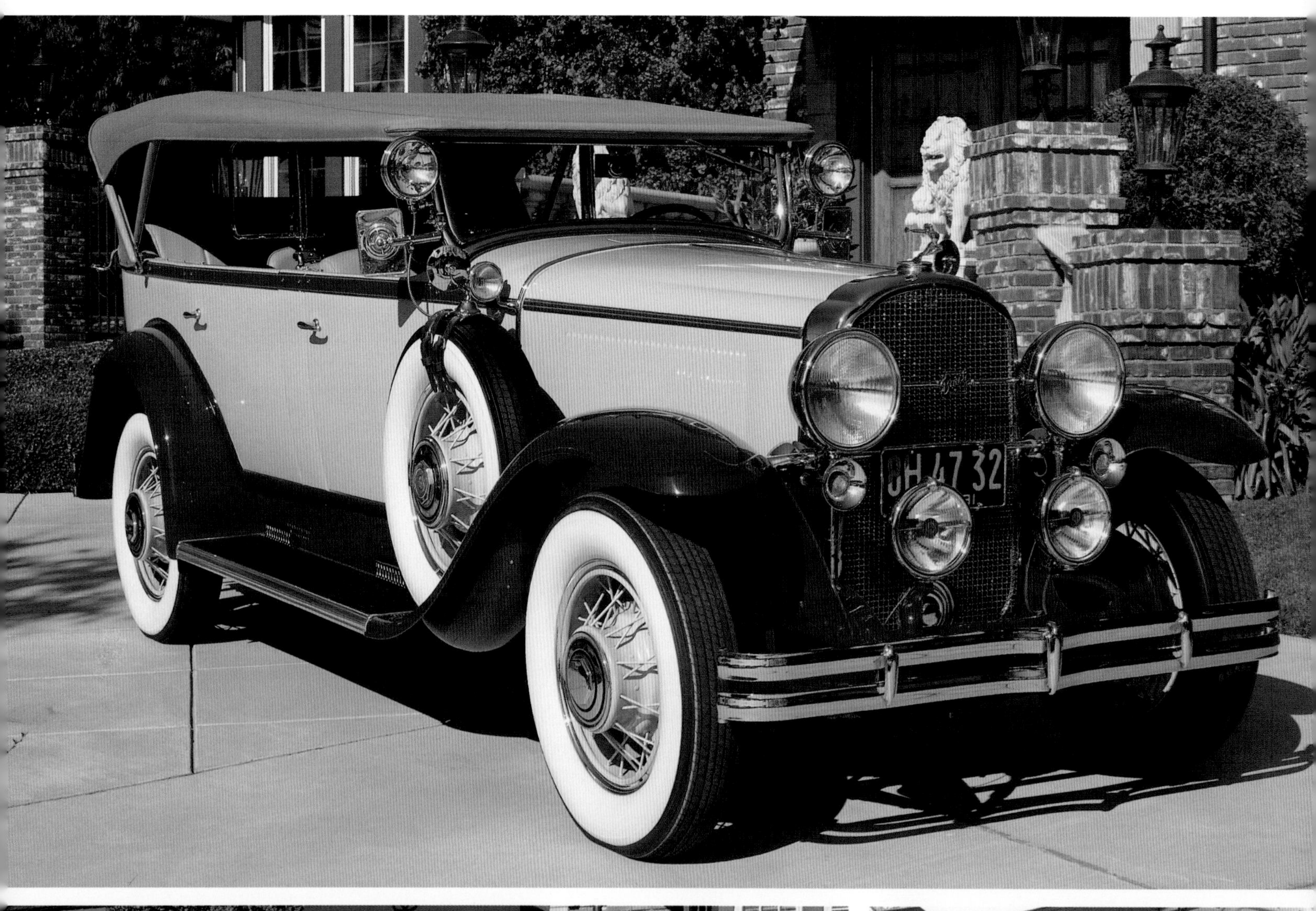
8H 47 32

8H 47 32
CAL 31

Buick

Nineteen thirty-one turned out to be a defining year for Buick. The marque was famous for decades for powering its cars with straight-eight engines. It was in '31 that this practice began.

The two sizes of sixes offered in 1930 were replaced by a trio of ohv (or "valve-in-head" as the division would refer to them) eights. The entry-level Series 50 was granted a 221-cid engine of 77 bhp, the step-up Series 60 featured a 90-bhp 273-cube powerplant, and the premium Series 80 and 90 cars came with a 345-cid engine good for 104 bhp.

The three smooth and reliable five-main-bearing eights were among the most advanced engines of their day. In late 1929, the division's chief engineer, Ferdinand "Dutch" Bower, placed development of the Buick eight in the hands of a team headed by 27-year-old John Dolza. The engines featured updraft carburetors, "V" belts to drive the fan, and aluminum oil pans. An automatic spark advance was a new feature at Buick, doing away with the need for a spark-setting lever on the steering column.

The year also marked Buick's adoption of synchromesh for its three-speed transmission. The improved gearbox was standard on all but the Series 50 cars.

Styling was little-changed from 1930, including a hint of 1929's controversial "pregnant" flare high up on the grille and hood. Each 1931 Buick series had its own wheelbase: 114 inches for the 50, 118 for the 60, 124 for the 80, and 132 for the Series 90.

When the decision to go to eights was made, the stock market crash of 1929 had not yet happened. By the time the cars reached the market, however, the Depression was on and Buick sales were down, slipping a place to fourth in the industry behind Plymouth in '31.

The Series 90 featured five- and seven-passenger sedans, a limousine,

two- and five-seat coupes, a roadster, a convertible coupe, and a seven-passenger phaeton at prices ranging from $1610 to $2035. The $1620 phaeton proved to be the rarest Buick of the year. It is one of just 460 made, 392 for domestic sale and 68 more for export.

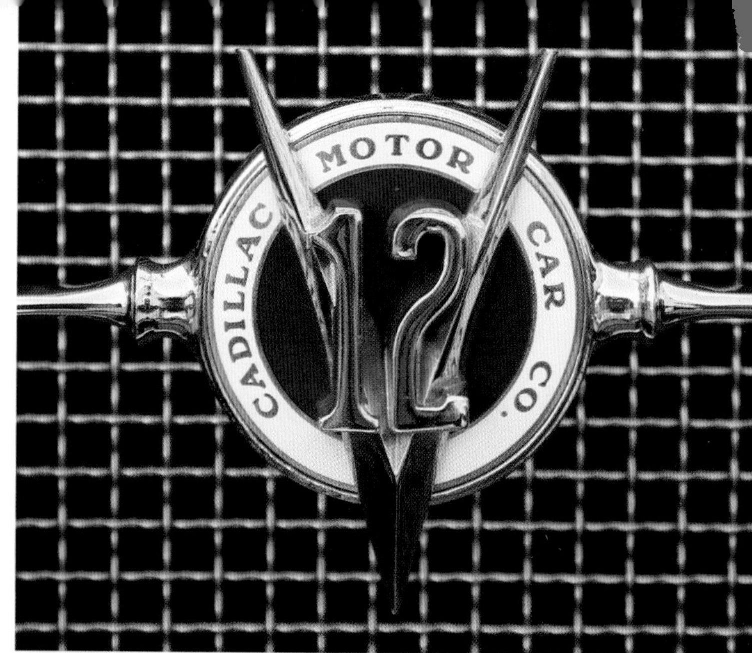

# 1931 CADILLAC
## 370A COUPE

The Cadillac V-12 will forever stand in the shadow of the marque's landmark V-16. The latter stunned the automotive world when introduced and continues to be favored over the V-12 with collectors. In its day, though, the Series 370 Twelve quickly outsold its more august brother by a considerable margin.

Cadillac's flagship V-16 made its debut on December 27, 1929. The V-12 followed in October 1930. With the 370, Cadillac had thorough coverage of the luxury-car market from its companion-car LaSalle starting at $2195 on up to Series 452 V-16s that could cost as much as $15,000 with custom coachwork. Twelves began at $3795 and tended to run about $2000 less than comparable Sixteens.

The V-12 engine was basically the V-16 minus four cylinders. The V-16 had overhead-valves and introduced the first hydraulic valve lifters to ensure quiet operation of the valve gear. Each bank of cylinders had its own carburetor. The V-12 inherited these features. Just as the V-16 operated as two straight eights on a single crankshaft, the V-12 was a pair of in-line sixes. The V-12 had a displacement of 368 cubic inches, and developed 135 bhp with 284 pound-feet of torque. Of course, it was less powerful than the larger V-16, but was nearly as smooth and was said to rev better at high rpm—perhaps helped by the shorter crankshaft. A lightweight V-12 roadster like the one that paced the 1931 Indianapolis 500 was good for around 85 mph, and even cars with heavier bodies could cruise at 60 to

HOLLYWOOD
6K 73 34
CAL 31

K 73 34
CAL 31

6K 73 34

CHOKE

70 mph. (Seven-passenger 1930-31 models were mounted on a 143-inch wheelbase that was three inches longer than that of other Series 370 cars.)

The V-16 and V-12 engines were not only well-engineered, but were also aesthetic triumphs. Raising the hood revealed a neatly laid out motor that was finished in black enamel, polished aluminum, and chrome.

In 1931, the coupe shown here was the least expensive Series 370A model. (Cadillac considered virtually identical cars produced during 1930 and 1931 to be part of the Series 370 and combined production numbers for the two years.) This 5035-pound vehicle was listed as a two-passenger car, but included a roomy rumble seat for another two riders. This car is said to be one of only four '31 370A coupes known to remain from a total production of 302.

ALT 753

# 1933 DELAGE
## D8S COUPE

The machines produced by Louis Delage exuded tasteful flair and quality. Born in 1874 at Cognac, France, he was a graduate in engineering from the respected School of Mechanical Arts in Angers. After two years as chief draftsman for Peugeot, Delage started his own carmaking firm in 1905, prototypes rolling from a barn containing four lathes, a drill press, and a milling machine.

A trio of Delages with single-cylinder De Dion engines were displayed at the Paris Auto Show in December 1905. They garnered accolades from a leading auto writer, prompting an investor to loan Delage 150,000 francs, providing his son be given a job.

By 1912, 350 employees turned out 1000 cars with reputations enhanced by racing success. Delages came in first and third at the 1911 Boulogne race, won the 1913 Grand Prix at Le Mans, and captured the 1914 Indianapolis 500. Profits led to a new factory at Courbevoie (in time to produce munitions for World War I) and enabled Delage to adopt the genteel lifestyle of his customers; a chateau at Le Pecq, a villa at Saint-Briac, a town house in Paris (where he maintained a showroom), and a yacht, *L'Oasis*.

A 10.6-liter Delage V-12 set the world land speed record in 1923 at 143.309 mph, and the jewel-like 1.5-liter dohc supercharged straight eights won every major 1927 Grand Prix race. The first of Delage's 4.0-liter, 105-bhp ohv straight-eight passenger cars debuted at the 1929 Paris Salon. If less exotic than the GP engines, the D8S had the low-rev torque to launch two tons of Gallic luxe over European roads.

Barker, Capron, Figoni, Fernandez et Darrin, Labourdette, Letourneur et Marchand, Pourtout, Saoutchik, Vanden Plas, and other coachbuilders cloaked low-slung Delages. But none were slinkier than this Freestone & Webb '33 D8S (Sport) coupe.

The well-crafted coupe stands only 59 inches high, gracefully complementing the 130-inch wheelbase. Drivers sit low in the sunken floor, their legs nearly straight as in a sports car. The back seat is intimate, but two couples and

judiciously selected luggage can be accommodated.

The richly veneered dash holds Jaeger gauges for oil temperature, oil pressure, amps, fuel, and a 4000-rpm tachometer. But water temperature is registered by the Boyce MotoMeter that sits six feet away on the radiator. With Marchal headlights one foot in diameter, 7.00 ×18 tires, pleasing proportions, a compact yet comfortable cabin, a dropped frame, and firm semi-elliptic springs fore and aft, the 120-bhp aluminum-bodied D8S looks and handles like a smaller car.

Push the lever of the well-wrought four-speed gearbox into first, and with a throaty rasp from the big, single-barrel Smith-Barraquand carburetor, the long, low automobile laughs at the years.

2C 75 91
19 CALIFORNIA 33

# 1933 DUESENBERG
## MODEL J CONVERTIBLE COUPE

"The ideal personal car for any man or woman. With top down it is the raciest of roadsters; with top up the sportiest of closed cars." So read the original Duesenberg brochure to describe the convertible coupe by Murphy.

The Murphy convertible coupe was Duesenberg's most-popular body style, with 60 built. It was also the least expensive at $13,500.

The Walter M. Murphy Company, of Pasadena, California, was an auto dealership that branched out into coachbuilding in the Twenties. Californian tastes were less conservative than those in the rest of the country, and Murphy catered to its market. The firm had several talented designers on its staff, particularly Franklin Hershey, who designed the first Ford Thunderbird; Phil Wright, who created the Pierce Silver Arrow for Chicago's "Century of Progress" World's Fair; and Frank Spring, who would head Hudson's styling department.

Duesenberg, meanwhile, set out to build the world's most powerful and opulent cars. It was natural that it would tap Murphy to supply custom bodies.

Although built in a series, each Murphy convertible coupe was different. Bodies were supplied to the Duesenberg factory without paint or trim. The customer could select paint, upholstery, interior hardware, and make minor variations.

Though Murphy closed in 1932, the first owner of our featured car—who bought it as a chassis—arranged to have a Murphy convertible body mounted.

The Duesenberg J was the most powerful American car before World War II. Its 420-cid straight eight with dual-overhead cams and four valves per cylinder put out 265 horsepower. In 1929, the first year of Model J production, 125 horsepower was about the most that could be had in an American car. Top speed for a light-bodied Duesenberg J was 116 mph.

Monitoring the magnificent engine was an impressive instrument panel that included the usual gauges plus tachometer, stopwatch, altimeter, and brake-pressure gauge. A mechanical "computer" activated the automatic chassis lubrication every 75 miles, and a dashboard light confirmed it. Warning lights signaled the need for oil changes and battery maintenance.

Duesenberg won the Indianapolis 500 three times, and chief engineer Fred Duesenberg was always looking for more power. In 1932, the SJ added a supercharger that raised horsepower to 320 and top speed to 130. The SJ was identified by chrome external exhaust pipes. Consistent with Duesen-

berg's policy of satisfying the client, the external exhaust could be ordered on naturally aspirated Js or added to older cars. Some SJ customers had the supercharger removed, but left the pipes.

115 SOUTH

# 1935 PACKARD
## EIGHT CLUB SEDAN

While the One Twenty, Packard's first medium-priced car, was the focus of attention in 1935, the senior models quietly carried on the marque's tradition of quality and luxury.

By the time of the stock market crash, half the luxury cars sold in the world were Packards. Perhaps no American car enjoyed as much respect in Europe and at home as the prewar Packard.

Packard policy dictated that styling changes be gradual so the previous model didn't look dated and resale value was maintained. The '35s were (for Packard) a major step forward into the streamlined age with more rounded bodies and "pontoon" fenders. Still, the cars were dignified and unmistakably Packards.

There were few mechanical changes from the previous model. Chassis were two inches shorter, but still retained dash-controled adjustable shock absorbers, power brakes, and automatic lubrication. A vacuum-operated pump shot grease to various lubrication points—eliminating the need for trips to a service station.

The senior models came in three series: Twelves, Super Eights, and Eights. The Eight Club Sedan pictured here was powered by a 320-cid L-head straight-eight good for 130 bhp—10 more than the previous year thanks to an aluminum head with higher compression. Packard's thorough engineering and demand for reliability was demonstrated by the engine's dual fan belts.

Such quality details added to production costs. By 1938, the senior models accounted for eight percent of production, but required the talents of half the labor force. Inevitably, the big Packards

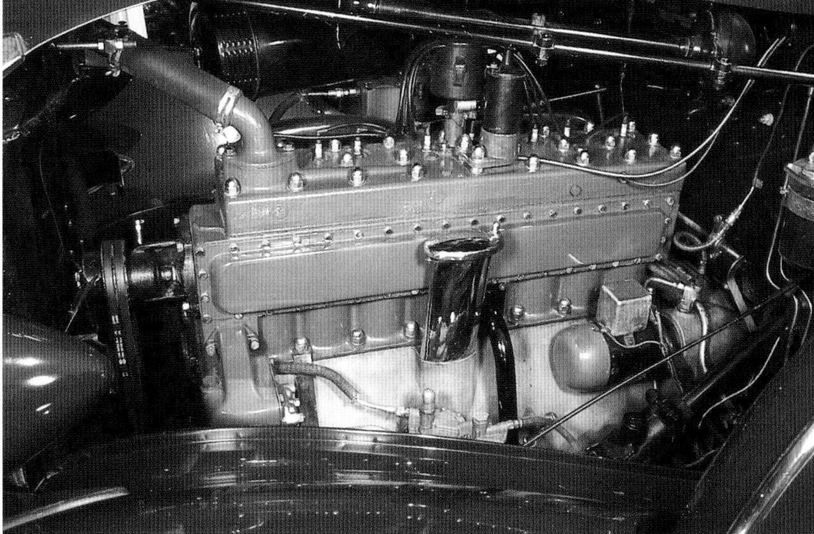

became more like the One Twenty.

That was still in the future when our featured car was built. Eights came in three chassis lengths; the Club Sedan was part of the 1201 range that rode on a 134-inch wheelbase. With its close-coupled styling, blind rear quarters, and built-in trunk, the Club Sedan was Packard's best-looking four-door. Base price was $2580. This car's appearance was enhanced by optional side-mount spare tires, Trippe lights, and a trunk rack. The radio was also optional, but Packard was the first to make a space for it in the instrument cluster.

2K 45 58
19 CALIFORNIA 36

# 1936 PACKARD
## ONE TWENTY LEBARON CONVERTIBLE VICTORIA

Packard was on the ropes in 1934, but by '36 it was paying healthy dividends. The reason was the medium-price One Twenty.

The luxury car market evaporated in the Great Depression and Packard needed an affordable mass-produced car to survive. Prices for "junior" One Twentys started at less than $1000, while the "senior" Packards started at well more than $2000. Thousands who had always dreamed of a Packard from afar could now afford one, and 10,000 orders were placed before car was even in the showrooms. Twenty-five thousand were sold during introductory model year 1935.

The One Twenty looked like a Packard with ox-yoke grille, red hexagons on the hubcaps, and dignified styling. Interiors were upholstered in wool broadcloth or leather. Built on a generous 120-inch wheelbase (hence the name), it was powered by a smooth-running straight eight putting out 110 bhp. Horsepower increased to 120 for '36—good for a top speed above 90 mph.

A modern assembly line was installed for the One Twenty. In 1936, half of Packard's work force crafted only

5985 senior Packards, while the other half assembled 55,042 One Twenties. In spite of mass production, the One Twenty was a well-built car with quality that was never an embarrassment to the company's lofty reputation.

The car shown here has a one-of-a-kind custom Victoria convertible body by LeBaron. The Victoria was a factory bodystyle for senior Packards but was not available for One Twentys. LeBaron went the factory Victoria one better by having a three-position top, which offered the option of driving with the roof erected over the back seat only. While factory bodies were steel, the LeBaron was aluminum. Most Packard custom bodies were mounted on a senior chassis, but the president of American Tobacco chose a One Twenty chassis. By choosing the junior chassis he got independent front suspension and hydraulic brakes a year before they were available on senior Packards.

This car was later owned by actor Edward Herrman, and its restoration was the subject of a History Channel show.

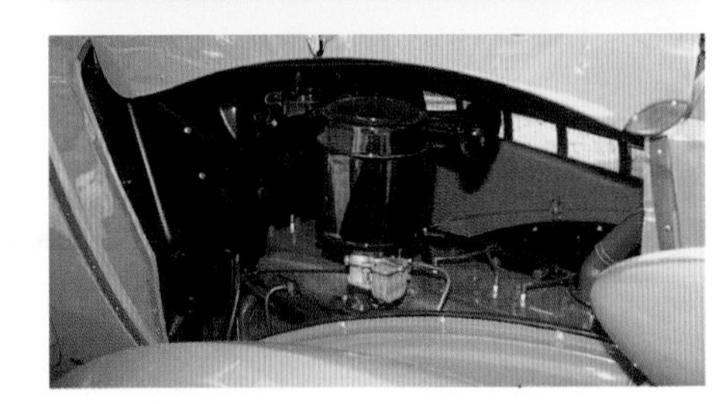

PACKARD
MOTOR CAR CO.

# 1937 CHRYSLER

## AIRFLOW FOUR-DOOR SEDAN

When it debuted in 1934, the Airflow was Chrysler's attempt to change the course of automobile design and engineering in the Thirties. Perhaps the most curious part of the Airflow story was that the normally canny Walter Chrysler approved the daring concept without much regard for whether the public would like it. As it played out, the design was successful in nearly every way but sales.

While the Airflow's smooth and rounded aerodynamic shape increased top speed and fuel economy, many car buyers of the day didn't find it as attractive as its boxy competitors. In fact, the styling was polarizing and often resulted in "love it or leave it" reactions.

The body shells were engineered along aircraft principles. This practice resulted in improved strength at less weight.

W.P.C. CLUB
CALIFORNIA
AERFLOW

W.P.C. CLUB
AERFLOW

Another of the Airflow's innovations was moving the engine forward so it was placed over the front axle. The increased room for passengers also allowed the back seat to be located ahead of the rear axle. This change in weight distribution resulted in a better ride—particularly for the rear-seat passengers who no longer sat directly above the pounding rear axle.

On the open road, Airflows cruised comfortably at 65 to 70 mph thanks to Chrysler's early adoption of overdrive. The speedometer featured a built-in tachometer, and a window at the bottom of the dial displayed engine rpms in third gear. When the automatic overdrive engaged, a window at the top of the dial took over. The feature wasn't particularly useful for driving, but Chrysler used it to demonstrate the drop in rpms with overdrive engaged along with the resulting decrease in fuel consumption and engine noise. The Airflow's streamlining also aided quiet cruising.

Power was provided by a 323.5-cid straight eight rated at 130 bhp. An automatic choke aided starting, but if the carburetor flooded, the driver could simply push a small button on the dashboard above the radio to drain the excess gasoline from the carb. The radio was optional, as was the dealer-installed heater.

The Airflow shown here is an end-of-the-road 1937 four-door sedan. It originally priced from $1610 and saw a production run of 4370 units. The only other Airflow model that Chrysler offered that year was the six-passenger coupe. It cost the same but was chosen by only 230 buyers.

The Chrysler Airflow started trends that we take for granted today. The design's unusual appearance and early production problems kept buyers away, but the eventual widespread adoption of some of the Airflow's pioneering principles proved its success in the long run.

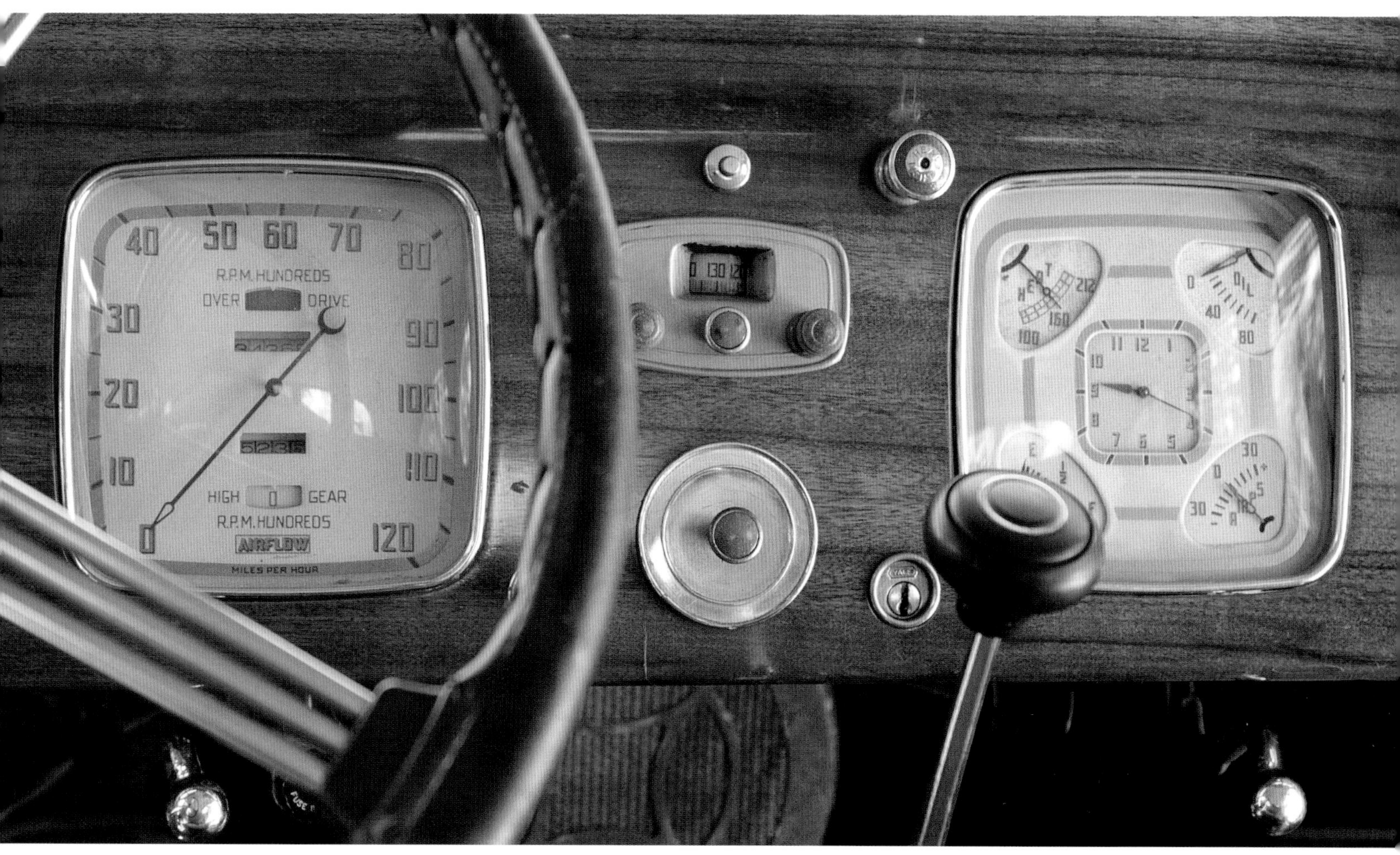
40 50 60 70 80
R.P.M. HUNDREDS
OVER DRIVE
30 90
20 100
10 110
HIGH GEAR
R.P.M. HUNDREDS
0 AIRFLOW 120
MILES PER HOUR
HEAT
212
160
100
OIL
0
40
80
11 12 1
10 2
9 3
8 4
7 6 5
E
½
F
30
0
30

CALIFORNIA
P111

# 1937
# ROLLS-ROYCE
## PHANTOM III TOURING LIMOUSINE

Writing about the Rolls-Royce Phantom III in *Automobile Quarterly* in 1979, Stan Grayson described it as "the most innovative and mechanically lavish of prewar Rolls-Royce cars." It would be difficult to argue the point.

Where the Phantom II had been powered by a six-cylinder engine, the "PIII" joined the growing circle of exclusive cars around the world with V-12s. Though a beam front axle had been good enough for its predecessor, the Phantom III was graced with a double-wishbone independent suspension. A Bijur centralized lubrication system served numerous points on the chassis. Built-in hydraulic jacks "made it possible to change a tire without requiring the passengers to disembark," as Grayson pointed out. The brakes derived a power assist from a servo driven by the transmission.

The new engine was a 60-degree undersquare design that displaced 448 cubic inches. Its block and heads were made of aluminum, and the overhead valves were hydraulically actuated. Unless the oil was changed often, the hydraulic valve lifters gave trouble and there was a switch to mechanical lifters in 1938. Breathing through a two-barrel carburetor and firing each cylinder with twin spark plugs, the Phantom III engine generated an estimated 180 bhp at 3650 rpm. Top speed for lighter-bodied cars was around 100 mph with a cruising speed of 80.

The front suspension was inspired by Cadillac's then-new design. However, Rolls-Royce employed horizontally mounted helical springs and hydraulic shock absorbers encased in oil-filled housings. Shock damping could be controlled from settings on the steering wheel hub.

First shown at the Olympia Motor Show in London in late 1935, the Phantom III was produced until July 1939. Only 710 were built before the World War II ended production. Of the 70 sent to North America, 18 received

wood-framed, alloy-skinned bodies by J. S. Inskip, Rolls-Royce's distributor in New York. A complete car could leave the Inskip shops priced a little bit on either side of $20,000.

The mostly original car seen here is one of the Inskip-bodied PIIIs. Classified as a "five-passenger sedan," it nonetheless includes twin divider windows and two folding jumpseats.

P111

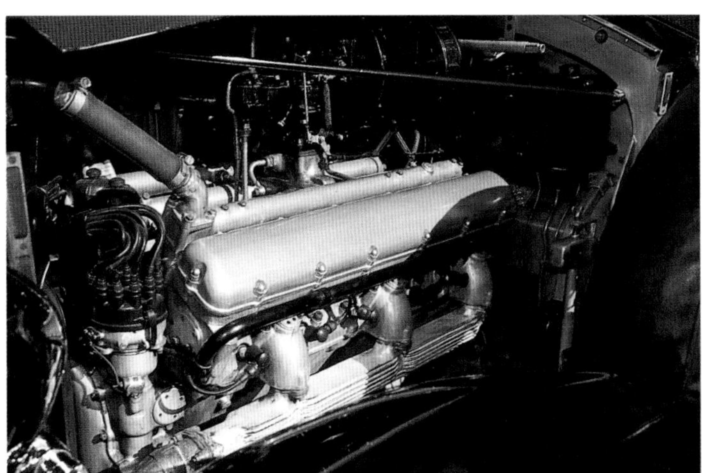

CALIFORNIA
P111

# 1938 CADILLAC

## SIXTY-SPECIAL FOUR-DOOR SEDAN

Bill Mitchell had been with General Motors' Art & Colour Section less than a year when he was promoted to head Cadillac and LaSalle styling. The Cadillac Sixty-Special was his first project and it proved the wisdom of his promotion. It was one of the great designs of Thirties and a milestone for Cadillac styling.

It has been estimated that Mitchell, who would later head General Motors's design department in the Sixties and Seventies, determined the appearance of 72 million cars. These included such clean designs as the 1963 Buick Riviera, '63 Corvette Stingray, and '66 Oldsmobile Toronado. But the '38 Sixty-Special remained one of Mitchell's favorite designs.

The Sixty-Special started out as a LaSalle sport sedan. Midway through the project, GM realized the car would be too expensive for a LaSalle and it became a Cadillac. Larger and more expensive than Cadillac's entry-level Series 60, but cheaper than the less-flamboyant large Cadillacs, the Special's price was $2090.

Built on a double-drop frame, it was three inches lower than other Cadillacs. This low profile improved handling and allowed for the elimination of running boards—a first for a car from the Big Three.

CAL 1938
71N1
Cadillac

The Sixty-Special was the first sedan with a fully integrated trunk. Previously, sedan trunks were upright boxes grafted on to the rear of the car. Plus, the Sixty-Special's fenders stretched rearward to make the trunk seem even more a part of the car, a look that is still with us today.

The distinctive roof looked more like a convertible top than a Thirties sedan roof. The side windows were large with thin, elegant chrome frames instead of the normal thick, painted stampings. The effect was similar to that of the postwar hardtop. Chrome trim was kept to a minimum. The car was sporty, daring, yet elegant. Cadillac brass was worried that it might be too radical for the ultra-conservative American luxury market. Their fears were unfounded; the Sixty-Special outsold every other Cadillac line in its first year. Sixty-Special features were adapted to other General Motors cars and soon copied by competitors.

Power was provided by an L-head V-8 putting out 135 bhp. This smooth and durable engine propelled Cadillacs from 1936 through 1948 (and it served reliably in tanks during World War II). Power was transmitted by a tough, quick-shifting three-speed gearbox that was later a favorite of hot-rodders. For '38, all Cadillacs moved the shifter from the floor to the steering column.

Cadillac would go on to dominate the postwar luxury-car market, aided by its influential overhead-valve V-8 engines and tailfins. Part of the dominance can be traced back to the design leadership that the Sixty-Special helped establish.

CAL 1938
71N1

# 1938 PEUGEOT

## 402 B RETRACTABLE HARDTOP

Chrysler's streamlined 1934 Airflow flopped in America, but it was an inspiration to French automaker Peugeot. With headquarters in Paris and its main plant in Sochaux near the Swiss border, Peugeot is the second-oldest automaker in the world.

In late '35, Peugeot introduced its tribute to Airflow styling with the model 402. The 402's rounded front with waterfall grille was the sincerest form of flattery. Peugeot took streamlining one step further than Chrysler by eliminating running boards and mounting the headlights behind the grille. Peugeot's streamlining was more than fashion, and the reduced wind resistance improved fuel economy and performance.

Capping the streamlined grille, Peugeot's lion's head mascot served not only as an ornament, but also as the hood latch. A second stylized lion graced the rear fender skirts.

Most 402s were sedans, but coupes and convertibles were also offered. The most interesting bodystyle was the Éclipse, which was a retractable hardtop. The metal top flipped under the trunklid for open-air motoring, yet offered the security and weather

5553
WE06

5553
WE06

protection of a coupe when raised. Tops on early models were lowered electrically, but starting in '37, a simpler mechanical system that could be operated by one person was used. Peugeot's retractable hardtop was unique in the Thirties. Ford tried the concept again in the Fifties with the Skyliner, but it wasn't until recently that retractable hardtops found commercial success.

Although styling was futuristic, engineering was conventional and lived up to Peugeot's tradition of tough, dependable cars. The Éclipse needed a lengthy 130-inch wheelbase to accommodate a rear deck long enough to swallow the retractable hardtop and to provide room for six passengers on wide bench seats. Independent front suspension contributed to good ride and handling. Power was provided by a 2.1-liter ohv four-cylinder that developed 63 bhp. The standard transmission was a conventional three-speed, but the shifter sprouted through the dashboard instead of the floor.

Fewer than 500 Éclipses were built before production ended in 1940.

# 1939 BUICK
## CENTURY CONVERTIBLE COUPE

Century first appeared on Buick's roster in 1936, the same year that Buick replaced its numeric series names with something more memorable: Special, Century, Roadmaster, and Limited. Century was most meaningful of the new names because it signified the car's top speed.

Often called "the first muscle car" or "the banker's hot rod," Century combined the smaller Special body with the 120-bhp straight eight from the bigger Roadmaster and Limited. The result was an excellent power-to-weight ratio giving a top speed in the vicinity of 100 mph and lively acceleration. At about half the price of an Auburn speedster, the Century was a performance bargain.

By '39, the 320-cid eight was putting out 141 horsepower. That was one more than Cadillac's most powerful V-8, a fact resented by Cadillac, which thought Buick was getting above its General Motors station.

The owner of this '39 Century doesn't hesitate to drive it in traffic because of its good performance, handling, and brakes. It is one of only 850 Century convertible coupes built for '39 at a base price of $1343. This car has several rare options such as sidemount spares and "streamboards," which replaced standard running boards for a more streamlined look. Unfortunately,

B 16 83

streamboards were delicate and few have survived. A much more popular option was the heater, which was mounted under the dash, near the passenger's feet.

Buick had several firsts in 1939. The most significant was the industry's first standard turn-signal lights. Unlike modern turn signals, these were not incorporated into the taillights, but were part of the trunk medallion. Buick also offered the first pushbutton radio tuning that year. New for Buick (but not the industry) was a column shifter that left the front floor unobstructed. The convertible coupe lost its rumble seat and replaced it with interior opera seats behind the front seat.

After almost 70 years in the Buick lineup (although not continuously), the Century nameplate was replaced by LaCrosse for 2004.

# 1940 PACKARD
## ONE EIGHTY DARRIN VICTORIA

There was no place for Dutch Darrin to go but home. After 15 glorious years of designing exclusive full-custom automobile bodies in Paris—first in partnership with fellow American Thomas Hibbard, then with a banker named Fernandez—he sensed the day of such highly individualized motor cars was passing. So, in 1937, Howard Addison Darrin, born 40 years earlier in Cranford, New Jersey, forsook the "City of Lights" for the "City of Angels."

Darrin set up in Hollywood, there to fashion customized variations of production cars he felt would appeal primarily to the free-spending film community. One of his early ideas was to cut the roof off of a "junior" Packard Eight/One Twenty coupe, replace it with a convertible top, and lower the hoodline. This

100
200

100
200

4PNE684

he did several times for a client list that would have made for quite a fine credit roll at the end of a Tinseltown blockbuster. Then the Packard Motor Car Company, that proper bastion of luxury-class motoring located in hard-working Detroit, got wind of what Dutch Darrin was doing in sun-drenched, fun-loving California.

Packard management elected to give its sanction to Darrin's efforts, carrying the custom models (Packard added a four-door convertible and a notchback four-door sedan) in its 1940 catalog and placing a chrome script of Darrin's signature on the cars. But there were conditions. Dutch favored the One Twenty for being light and easy to modify, but the company wanted him to build most of the cars on the Custom Super Eight (or One Eighty) chassis. The $4593 Convertible Victoria, as the two-door soft top was billed, rode a 127-inch wheelbase.

Aside from sectioning the hood and radiator shell to lower its profile, the Victoria had cut-down doors. Despite the "roadster" look of its thin windshield frame, it was a true convertible with roll-up side windows. The Darrins were the first Packards without running boards. Initially built in the former Auburn body plant in Connersville, Indiana, production later shifted to Ohio funeral-car builder Sayers & Scoville. Victorias were made until 1942.

This restored 1940 Darrin Victoria is powered by a 356-cid L-head straight-eight engine. The powerplant's 160 bhp are driven through a three-speed transmission with overdrive.

Darrin

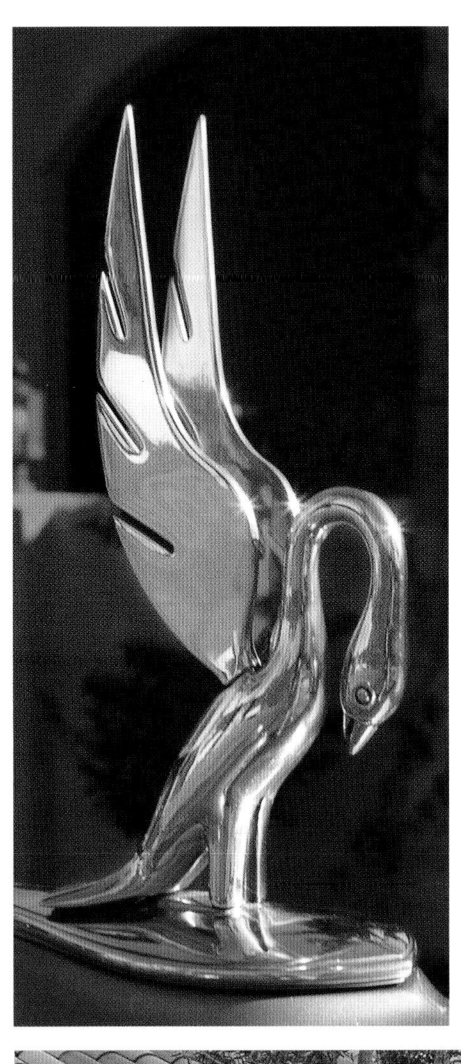

# 1940 PACKARD

## ONE EIGHTY DARRIN FOUR-DOOR CONVERTIBLE

Packard, the conservative luxury-car maker, survived the Great Depression largely by moving down into the middle of the automobile market where it could sell cars in higher volume. By the late Thirties, though, the firm's preeminent engineering and long-held practice of stylistic evolution was no longer enough to prosper in the high end of the market.

More aggressive competitors, particularly Cadillac, were succeeding in the quickly evolving marketplace with increasingly stylish and more modern designs than Packard offered. Packard's management was savvy enough to recognize this shortcoming and, by the end of the decade, it was taking steps to address the problem.

Howard "Dutch" Darrin, a Cranford, New Jersey, native, had designed full-custom car bodies for an exclusive clientele in Paris for a decade and a half before relocating to Southern California in 1937.

Stateside, Darrin aimed to sell production-based custom cars to Hollywood's free-spending motion-picture crowd. Dutch settled upon the medium-price Packard One Twenty as a base for his creations and found some success. The first handful of cars were built at Los Angeles-area body shops. Before long, however, Darrin had set up his own facility in an old bottling plant on the Sunset Strip in Hollywood.

Word of Darrin's restyled Packards soon made it to headquarters in Detroit. Then, after an unauthorized showing

during an event at the Packard Proving Grounds, dealers started pestering management to offer factory-authorized Packard Darrins through the company's sales channels.

Packard brass eventually agreed to catalog Darrins, but with a few stipulations. The majority of them had to be built on the more-prestigious Custom Super Eight chassis and they wanted to add a pair of four-door cars—a convertible sedan and a closed sport sedan—to Darrin's rakish convertible victoria.

Dutch agreed and quickly arranged for the former Auburn Automobile Company to produce the 1940 Packard Darrins at its underutilized Connersville, Indiana, body plant. This plant once built Auburns and Cords and would produce Jeep bodies during World War II. Prices ranged from about $3800 for the handful of Victorias that were built on the One Twenty chassis to around $6300 for a Custom Super Eight convertible sedan. Thus, the four-door ragtop was about $2000 pricier than the most-expensive Detroit-built Packard that year.

The two four-door cars were built on a 138-inch wheelbase and shared much of their custom body work. Supposedly all of their inner structure and skin, except for a section of the hood and the trunklid, were completely different from regular-production Packards. Power came from a 160-bhp 356-cid inline eight.

Most historians say only two of the sport sedans were built, along with about a dozen of the convertible sedans in 1940. As many as 50 of the Victorias may have been constructed that season, with about half on the 127-inch wheelbase used by some Custom Super Eights.

This car is thought to be one of nine 1940 Packard Darrin convertible sedans known to exist. The Packard Super Eight is a fine highway car that can cruise at 70 to 80 mph in overdrive.

# 1941 CADILLAC
## SERIES SIXTY-TWO CONVERTIBLE SEDAN

This 1941 Cadillac Series Sixty-Two has a tank engine under the hood. Using the tank engine wasn't an attempt to make a monster Caddy. During World War II, M-5 and M-24 light tanks were powered by twin Cadillac L-head V-8s driven through Hydra-Matic automatic transmissions. The powerful and reliable Cadillac engines were modified for tank duty and proved their worth in battle. General Motors' Hydra-Matic was the first fully automatic transmision and was introduced in Oldsmobiles for 1940. Hydra-Matic was new technology and prewar automatics weren't as durable as the Cadillac engines. Wartime development greatly improved their longevity and thanks to tank duty, postwar Hydra-Matics had a solid reputation.

Converting a Cadillac tank engine to passenger-car use meant replacing most of its components, but the final result was a smooth-running Cadillac—and a reliable one.

LaSalle, Cadillac's companion make, was discontinued after 1940 and

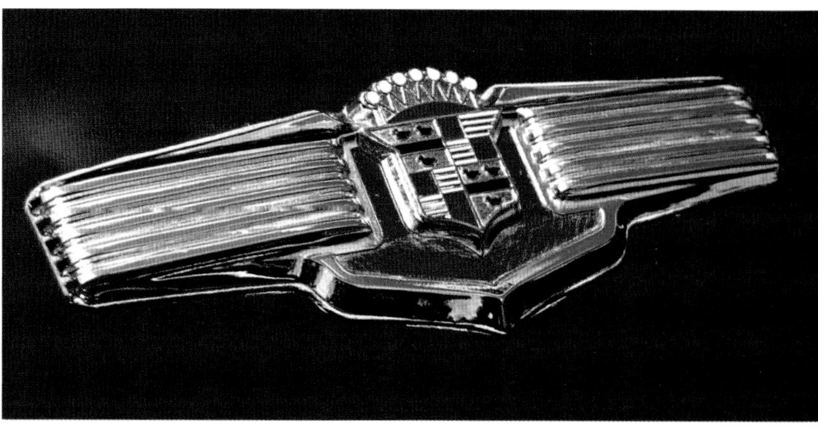

HISTORIC VEHICLE

replaced by a new entry-level Cadillac Series Sixty-One. The 1941 Series Sixty-Two was priced just above the Series Sixty-One. Sixty-Two prices started at $1420 for a coupe and the convertible sedan was the most expensive model at $1965. Although Cadillac two-door convertibles had a power top, the big top on convertible sedans was raised and lowered manually. In spite of its size, the well-engineered top could be handled by one man. Perhaps the four-door convertible didn't get a powered top because '41 would be the last year for the body style at Cadillac. Convertible sedans had their heyday in the Thirties and were offered by most manufacturers, but they were never big sellers. By the start of the Forties, popularity was waning and four-door ragtops were being dropped. Cadillac sold only 400 in '41, but that was good compared to the 122 sold in 1940.

The rest of the Cadillac line was doing well, and Cadillac had a record year in 1941 with 66,130 cars built. It was easy to see why Cadillac was successful that year. New frontal styling included a broad eggcrate grille that would be a Cadillac theme for decades to come. Also new were prominent taillights with the left-side reflector assembly concealing the gas cap. The 346-cid V-8 had higher compression and other improvements that upped horsepower to 150. Top speed for most models was more than 90 mph. The Hydra-Matic automatic transmission was a popular new option for 1941 Cadillacs, although the car on these pages used Cadillac's strong, yet smooth-shifting, three-speed manual.

4T1 CAD

# 1941 CHRYSLER

## NEWPORT DUAL-COWL PHAETON

LeBaron built some of the most beautiful custom bodies of the classic era. It's fitting that the coachbuilder would go out in a blaze of glory with two dazzling show cars.

Briggs Manufacturing, which bought LeBaron in the Twenties, built bodies for Chrysler. Briggs sold Chrysler on two "idea cars," but they had to be built in less than five months for the 1941 New York Auto Show in October 1940. That required the skills of LeBaron. These would be the last cars built by LeBaron (though the name would later appear on certain Chrysler models). The Thunder-

bolt was the more futuristic of the two with its full-envelope body and curved windshield, but it was the Newport that was the more glamorous.

The Chrysler Newport looked both to the past and to the future. Its dual-cowl body with two folding windshields was already an anachronism by the mid Thirties, while its streamlining was ahead of the times. In 1941 some carmakers were cautiously extending the fenders into the front doors, but the Newport's front fenders swept all the way back to the rear fenders. The hidden headlights first appeared on 1936-37 Cords, reappeared on '42 DeSotos, then disappeared from American cars until the 1963 Chevrolet Corvette.

Ralph Roberts, one of the first LeBaron employees, was responsible for the styling, and the LeBaron heritage shows in the rakish, flowing lines. An aluminum body not only saved weight, but it was easier to pound out in a hurry.

The Newport was built on a 145.5-inch-wheelbase Chrysler Crown Imperial chassis. Power was provided by the standard 143-bhp Imperial straight eight, and shift-reducing Fluid Drive was included.

The Newport and Thunderbolt were hits at the New York show. Five more copies of each design were built and sent to Chrysler dealers around the country to drum up showroom traffic. The Newport was also chosen as the

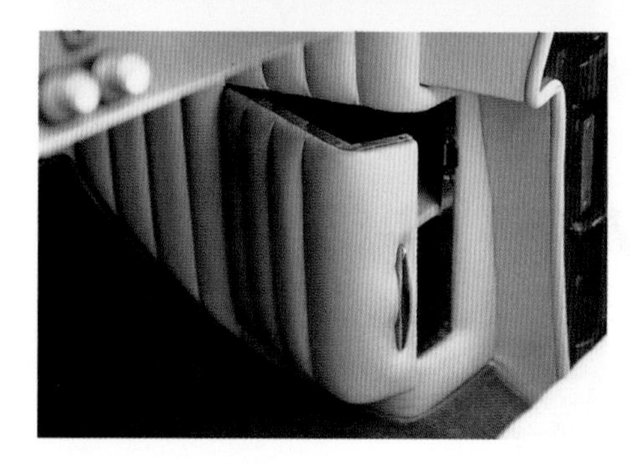

1941 Indianapolis 500 pace car.

After their tour of duty in Chrysler showrooms, five Newports were sold to the public at an estimated $6000 a copy; Walter P. Chrysler Jr. kept one for personal use. The one seen here was first owned by millionaire Henry "Bob" Topping. Topping was married to movie star Lana Turner from 1948 to 1952. Even though it isn't known how much Turner drove this Newport, this car is always associated with her.

Topping did some customizing on the car. The Chrysler engine and transmission were replaced by a Cadillac V-8 and Hydra-Matic transmission, probably just before or during World War II. Topping liked to put his name on things; thus, the Topping name appears on the engine heads, steering wheel, and hubcaps, and his initial is on the grille.

# 1941 PACKARD
## ONE EIGHTY LEBARON SPORT BROUGHAM

In 1899, James Ward Packard rode in the first horseless carriage of his own design. Packard had previously bought a Winton car. The car was troublesome and Packard decided he could build a better car—at the urging of an exasperated Alexander Winton. James and his brother, William Doud Packard, took their background in electrical equipment and machinery, and added Alexander Winton's former principal engineer and shop superintendent, laying the groundwork for an automotive dynasty that lasted until 1958.

When Packard released its Nineteenth Series models One Ten, One Twenty, One Sixty, and One Eighty in 1941, no one had to "Ask the man who owns one" what it meant to drive one. Changes in this series included headlights molded into the fenders, larger radiator side grilles, a longer hood, and fender-mounted parking lamps. The One Eighty was available on special order with bodies custom-built by independent coachbuilders, and LeBaron was one of the best. The LeBaron Sport Brougham was a new design for 1941. It was only offered on the 138-inch wheelbase of the topline One Eighty series. Buyers had a choice of painted or leather-covered roof. Power was provided by Packard's super smooth, 160-horsepower straight eight.

This One Eighty LeBaron Sport Brougham is thought to have been owned by a member of the Rockefeller Family. The Carmine Red Metallic car has a selective synchromesh three-speed transmission with overdrive and a Dual-Stream Heater with auxiliary

19 CALIFORNIA 41
38 T 234

CALIFORNIA
3BT 234

defroster. An option not ordered was air conditioning. Packard was the first to offer factory air conditioning in 1940. Most of the unit was mounted in the trunk and the air outlet was on the parcel shelf, behind the rear seat. The system didn't work very well and was discontinued after 1942.

This car is one of 99 special four-door sedan bodies commissioned from LeBaron; an estimated 26 are known to still exist. An original buyer paid $3545 for its distinctive side mounts, enameled hubcap emblems, and Cormorant radiator mascot. The running boards were a free option that could be ordered in place of the chrome gravel shields delivered with the car.

With its Safety Flex front suspension, the car cruises at 70 mph effortlessly and in total comfort.

# 1941 PACKARD
## ONE SIXTY DELUXE CONVERTIBLE COUPE

Packard's policy of gradual styling changes helped it to maintain a gold standard of resale value and allowed owners to keep their cars longer without looking dated. This linear styling policy served Packard well until the Forties. By then, though, American car design was changing at an incredible rate. Packard's unhurried design evolution couldn't keep up with the pace, and by '41, its cars looked old-fashioned.

But Packard wasn't out. Late in the 1941 model year, Packard brought out its highly acclaimed Clipper, which was

CALIFORNIA
9W 50 58

lower, wider, and more modern than the competition. Packard styling was once again esteemed. The company was in the process of replacing its old-style bodies with Clipper styling when World War II broke out. (The obsolete design did have an unexpected admirer in Soviet Premier Joseph Stalin. Russia's ZIS 110, made from 1946 to 1959, copied 1941-42 Packard styling.)

Although not well received in '41, to modern eyes, the Packard One Sixty Deluxe convertible coupe shown here looks like a classic example of a prewar convertible. Styling was subtly modern-

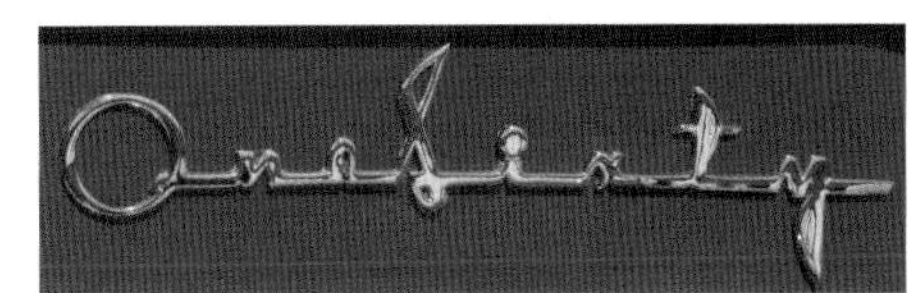

ized for 1941, with the headlamps integrated into the front fenders and capped with prominent chrome spears holding the parking lights at their tips.

Most makes had phased out sidemount spares by '41, but the upright styling of the One Sixty carries them well. Running boards were also on the way out—Packard had moved them to the options list.

Although One Sixtys were senior-model Packards, they shared body panels and a 127-inch wheelbase with the medium-price One Twenty (though there were some longer-wheelbase One Sixty sedans). They did have more upscale trim, and this was most evident in the interior. The convertible offered a choice of cloth and leather or full leather—as on this car. As a Deluxe model, the featured car has inlaid wood window trim. By the Forties, American luxury cars had replaced real wood with plastic moldings or woodgrain painted on metal, but the wood window trim on this car is as fine as in any custom body from Twenties or Thirties.

Another interesting trim detail in the interior involved the control knobs on either side of the steering column. To operate these controls, the driver grabbed rectangular handles that were shaped, and colored, like a Kit-Kat candy bar.

What really set the senior Packards apart from the One Twenty was their big 356-cid straight eight. New for 1940, this heavy engine was incredibly smooth and quiet. It had a sturdy 105-pound crankshaft running in nine main bearings, and was the first Packard with hydraulic valve lifters. The motor put out 160 bhp (10 more than Cadillac's V-8) and could push the two-ton convertible past 100 mph.

Packard built at least 128 One Sixty Deluxe convertible coupes for '41; only 19 are known to have survived. The DeLuxe drop top priced from $2112, while a base model listed from $1937.

# 1947 BENTLEY
## MARK VI DROPHEAD COUPE

Bentleys were fast sport tourers—absolutely dependable, but loud. By 1931, when Bentley Motors went into receivership, its larger cars were competing with Rolls-Royce. In a surprise move, Rolls bought its English competitor to prevent future rivalry.

The Rolls/Bentley used a modified version of a Rolls-Royce engine and a chassis from a shelved baby Rolls. Offering good performance and handling with Rolls-Royce refinement, Bentley was dubbed the "Silent Sports Car." The Mark V introduced a new chassis with independent front suspension just as Hitler was invading Poland. The postwar Mark VI used a revised version of that chassis. Precise steering and good handling did justice to the Bentley name, while ride was up to Rolls-Royce standards. One hundred thirty bhp was provided by a 4.25-liter F-head version of the prewar ohv six. Intake valves were overhead, while the exhaust valves were on the side. This allowed unusually large intake valves for better breathing. In 1951, Bentley enlarged the engine to 4.5-liters for 150 bhp, which increased top speed from 90 to more than 100 mph.

Before World War II, Rolls-Royce and Bentley only built chassis. Bodies were supplied by coachbuilders. Postwar survival demanded that Rolls and Bentley provide bodies as well. About 80 percent of Mark VI production left the works with a factory body. Monsieur Gudol, a French munitions dealer, preferred a custom body. He ordered this amazing car as a reaction against wartime austerity and to win Concours d'Elegance. This Bentley was first seen at the 1947 Paris Auto Show. Following its debut, the Bentley won two major French concours.

The cabriolet body by Franay combined styling elements from two other leading French coachbuilders: Saoutchik and Figoni et Falaschi. French coachbuilders of the late Thirties used a teardrop shape to achieve superior aerodynamics and produced some of the most flamboyant cars ever built. Front wheels were often enclosed in voluptuous skirted fenders. This increased the width

6101
9055 RQ

6101
9055 RQ

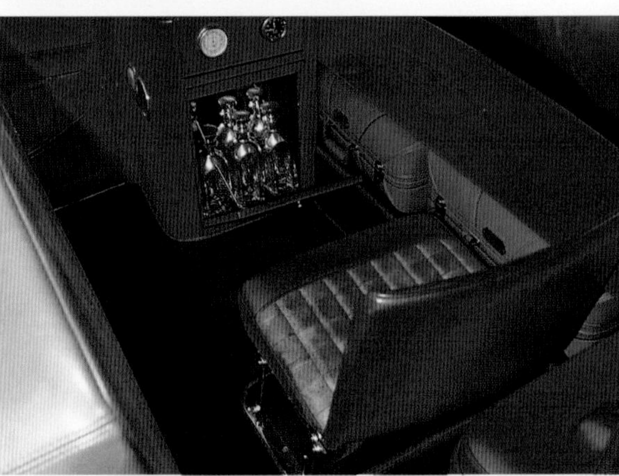

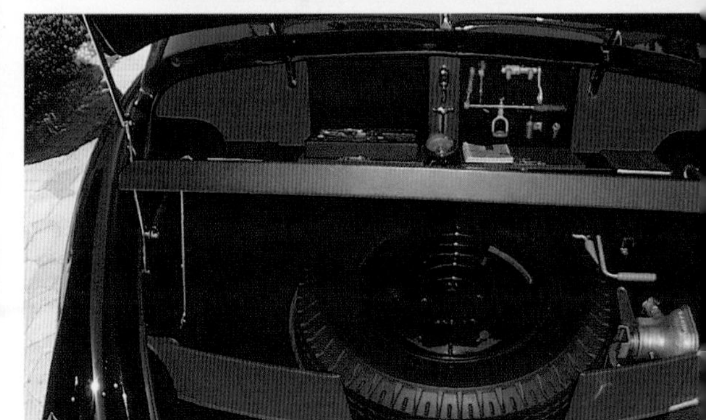

of the Franay Bentley to seven feet.

The sumptuous interior was as extravagant as the exterior. The driver was cossetted on a huge leather bench seat with a four-speed gear-change by his right knee. The interior was accented with frog hide. A rear passenger sat in a jump seat facing a full bar, his feet resting near Franay fitted luggage. In the event the Mark VI failed to live up to its reputation for reliability, the trunk contained a comprehensive collection of tools and manuals. In the early Fifties, Gudol had the car updated with Bentley's enlarged 4.5-liter engine. With its exotic bodywork, this car is possibly the most valuable Bentley in the world.

# 1947 STUDEBAKER
## GARDNER SPECIAL ROADSTER

Nearly 65 years after it was created by talented but mercurial designer Vince Gardner, the Studebaker Gardner Special roadster was still turning heads.

When Gardner began the Studebaker Gardner project, he was a modeler and stylist for Studebaker in South Bend, Indiana, and the starting point for the Special was his personal 1947 Champion business coupe.

Gardner cut the roof off the coupe; moved the front seat, pedals, cowl, and steering assembly back; and lengthened the hood and front fenders. He topped it with a Plexiglass roof that mated to a headerless windshield with aluminum pillars that were anchored to the reinforced chassis frame. For open-air motoring, the top could be removed and stored in the trunk.

The hood and decklid were lower than those of production cars and could be raised via hydraulic lifts that Gardner installed. He fashioned a single-bar grille with a central spinner and used '49 Studebaker Commander bumpers. Exhaust outlets were routed through the centers of the circular taillights.

Gardner called on his early experience as a Cord clay modeler to design an interior with a central gauge cluster and toggle control switches on the steering column. Tan cowhide covered the seats and interior door panels.

The engine was a Champion 169.6-cid L-head six to which Gardner added a Weiand twin-carburetor manifold and high-compression aluminum head, plus split exhaust manifolds. The transmission was a conventional Stude three-speed manual with overdrive.

In 1949, Gardner used the car to win a 450-mile rally conducted by the

Detroit chapter of the Sports Car Club of America. The following year, he took the roadster to California, entering it in the Oakland Roadster Show and the first *Motor Trend* magazine Autorama show in Los Angeles. Art Center School design students got a look at it at their LA campus, too.

Back home in Indiana, though, Gardner's wife—the couple would soon divorce—began prodding him to junk the car, which he did. However, the Special was saved from oblivion by a young man named W. Alan Canty Jr. (Canty later became a psychologist and had Gardner as a patient.) After Canty was murdered, his widow put the roadster up for sale. It wound up back in South Bend for many years, then passed to a Stockton, California, man.

The Gardner Special had many hard years before eventually passing through the doors of restorer Fran Roxas's shop in the Chicago area. (Roxas had restored another of Gardner's notable creations, the Ford-Vega roadster completed in 1953.) After two years of work, including undoing modifications that had been made to the car over time, the Gardner Special was ready for its star turn at the Pebble Beach Concours d'Elegance, where it claimed a second-place award in the "American Sport Customs" class.

1947
STUDEBAKER
PROTOTYPE ROADSTER
BY Vince Gardner

# 1947 TRIUMPH
## 1800 ROADSTER

Triumph might have died in 1939 if it hadn't been for Sir John Black's bad temper and Jaguar envy. Sir John was the unpredictable, but effective, manager of Standard Motor Company, Ltd. He was famous for wild mood swings.

Standard supplied engines for early Jaguars and at one time Sir John wanted to absorb the burgeoning Jaguar into Standard. William Lyons of Jaguar refused.

Sir John decided Standard wouldn't build Jaguar sixes after World War II but would sell Jaguar the tooling. Lyons knew to jump on the deal before Sir John changed his mind—which he did. Black was used to getting his way and took offense. He decided to beat Jaguar at its own game.

Before the war, Triumph made a name, but not much money, building upper-medium-priced cars with a sporting reputation and some success in rallies. With the Triumph name and Standard's resources, Black planned to put Jaguar out of business. Unfortunately by VE Day there wasn't much left of Triumph other than the name. What hadn't been sold off had been bombed by the Germans.

Jaguar unwittingly helped with the engine. Although Standard had stopped building Jaguar sixes, it still built Jaguar's 1800cc four. This engine started out as a Standard side-valve, but in the mid Thirties, Jaguar added a new cylinder head with overhead valves for increased power. The 63-bhp unit was the only suitable engine Standard had for its new sporting Triumphs. The four-cylinder Jaguar and Triumph also shared the same four-speed transmission, although Jaguar used a floor shift while Triumph's shift was on the column. The Triumph

BMS 154

California
5JNA465

Roadster was not a sports car, but it was sporting. Top speed was 77 mph, and 0-60 took 25.2 seconds. *Motor* described the Roadster as having "a distinctly better than average all-round performance." The '47 Volkswagen, for example, never made it to 60 but topped out at 57.3 mph in *Motor* tests.

The chassis used a Standard independent front suspension with a transverse leaf spring. Sheet steel was severely rationed in postwar England, so tubular steel was used for the frame. The price for a '47 Roadster was a rather steep £775; the cheapest Jaguar sedan with same engine cost £676.

Although the body was new, it had a prewar flavor with its classic chrome radiator and pontoon fenders. Open-air rumble seats faded away in the late Thirties, but Triumph had the distinction of building the last rumble seat. The last was perhaps the best because opening the seat raised a second windshield

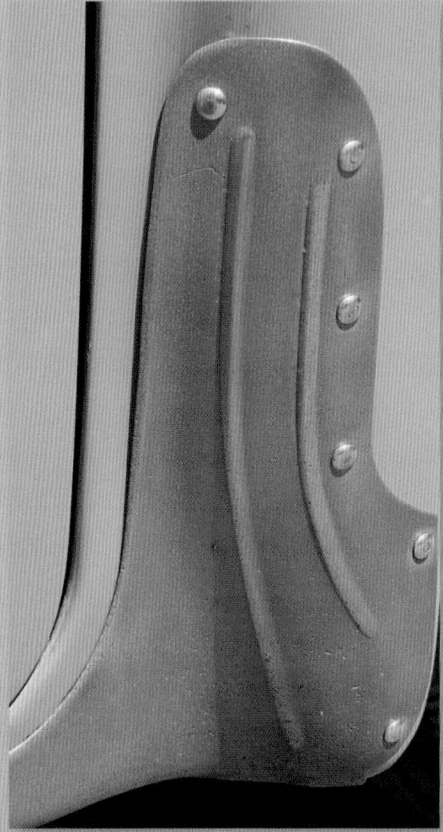

that offered some wind protection. The Roadster sat five, two in the rumble seat and three crammed in leather-and-wood luxury on the front bench seat. The Roadster was a modest success with 4500 built between 1946 and 1949. Triumph didn't put Jaguar out of business, but its name would supplant Standard.

# 1949 CADILLAC
## COUPE DE VILLE SHOW CAR

When the doors of New York's Waldorf Astoria Hotel opened on January 20, 1949, to admit visitors to the "Transportation Unlimited" exhibition, the public got a peek into the near-term future of the automobile, at least as General Motors saw it. That future was

going to include "hardtop convertibles," steel-roof coupes without fixed B-pillars that strove to capture the look of convertibles with their fabric tops raised.

All five GM car divisions showed a two-door hardtop model at this first of what would turn into the famous Mo-

torama shows. Before the '49 model year was over, Oldsmobile, Buick, and Cadillac would have two-door hardtops on the market. However, where the Olds and Buick hardtops in dealer showrooms looked nearly identical to the ones in the Waldorf ballroom, the Cadillacs were quite different.

The 2150 Cadillac Coupe de Villes made for public sale were part of the 126-inch-wheelbase Series 62 lineup. They bore a roof design like that of the Olds and Buick hardtops, which featured a wraparound backlight and rear pillars that narrowed toward the bottom. The show version was stretched over the

133-inch chassis of a Series 60 Special sedan and featured a more enveloping roof with smaller limousinelike back glass.

Other touches unique to the show car included a one-piece curved windshield, chrome wheel-opening trim, simulated air scoops on the rear fenders, and hydraulically operated side windows and vents. The custom interior was outfitted with a shortwave telephone and rear arm rests that held a vanity case, lipstick holder, and perfume atomizer.

Information compiled by the General Motors Heritage Center shows the one-off car was built up in part from a 1948 convertible. The roof was fabricated from three large longitudinal sections welded together, and there is wood bracing around the back window and other places.

After eight days at the Waldorf and another week at Convention Hall in Detroit, the car passed to GM President Charles Wilson. When he became U.S. Secretary of Defense in 1953, he gave it to his secretary, who took it with her to California. In 1957, the '49-vintage 331-cid ohv V-8 was replaced by a 365-cid Eldorado V-8 fed by dual carburetors and rated at 325 bhp. Power brakes and steering were added at this time, too.

ONTARIO
3H3192

MICHIGAN
49 BABY
GREAT LAKES 92

# 1949 LINCOLN
## CONVERTIBLE COUPE

The names Lincoln and Mercury had long been linked in corporate planning at Ford Motor Company. As products, though, the two brands were perhaps never closer than they were in 1949.

Ford's first all-new cars since before World War II came out for '49, including two lines of Lincolns. At the top sat the Cosmopolitan series, with four distinctive bodies. The base series shared its shells with Mercury, however. In fact, as originally conceived by stylist E. T. "Bob" Gregorie, the "baby" Lincoln was supposed to be the new postwar Mercury (and the Merc the new Ford). Then a rethinking by Ford's new management team boosted all the lines up a step, allowing for a line of smaller Fords to be created.

The entry-level '49 Lincoln came in three body styles: coupe, four-door sedan, and convertible. They shared the Mercury's rooflines, divided windshield, and front fenderline that faded into the doors near the body midpoint. However, "9EL"-series cars sported sunken headlights (glass covers were

planned), "frowning" grille, and circular taillights. These features visually linked the base models to the costlier Lincoln Cosmopolitans.

Then, too, both Lincoln branches shared a powertrain—and it was new. Gone were the troublesome V-12s of the Zephyr years, replaced by a 152-

bhp "flathead" V-8 displacing 336.7 cubic inches. It was driven through a three-speed manual transmission (though General Motors's Hydra-Matic automatic became a late-season option).

At 121 inches, the base Lincoln's wheelbase was four inches shorter than the Cosmo's—but three inches longer than Mercury's. All shared in the corporate shift to independent front suspension, semielliptic rear leaf springs, and open driveshafts.

Introduced in April 1948, the two Lincoln series accounted for more than 73,000 cars for the long model year. They were facelifted for 1950 (and again for '51), but the $3116 base-series convertible would not return after that first season.

The rare "baby" Lincoln ragtop seen here was outfitted with optional fog lamps and overdrive.

Chrysler
HISTORICAL
OHIO
110XIX
VEHICLE
54
CHRYSLER

# 1950 CHRYSLER NEW YORKER FOUR-DOOR SEDAN

A big boost to Chrysler's image was in the offing in 1950, an infusion of performance that would result from the arrival of the ohv "hemi" V-8 in 1951. Up to that point, though, Chrysler had carved out a good reputation for solid quality and riding comfort.

Underhood in those pre-V-8 days were proven L-head inline engines, a 250.6-cid six in Royals and Windsors and a 325.5-cube eight in the Saratoga, New Yorker, wood-trimmed Town & Country, and the Imperials. Features like "Hydra-Lizer" double-acting shock absorbers, "Cyclebonded" brake linings, and "Prestomatic Fluid Drive" transmission made Chryslers safe and comfortable to drive. If styling was less adventurous than that of some competitors' cars, company president K. T. Keller was more than happy to hang his hat on the generous head room and chair-height seats in Chrysler interiors.

"A car of comfort . . . a car of class . . . that's the Chrysler New Yorker," wrote *Motor Trend* Editor Walt Woron after testing a 1950 four-door sedan.

Though the Windsor was the breakaway overall sales leader for Chrysler in 1950, the New Yorker headed the eight-cylinder group, most of which was on a 131.5-inch-wheelbase chassis. The series was comprised of a club coupe, a convertible, a Newport two-door hardtop that was making its debut, and the $2758 sedan that accounted for 22,633 of the 29,335 New Yorkers made for the year.

Core bodies and chassis were the same as in 1949, when Chrysler ushered in its first completely new cars since the end of World War II. There were plenty of new touches to distinguish the '50s, however, starting with a simpler—but still bold—grille. Squared-off rear fenders imparted a sense of greater length and newly incorporated

Chrysler

the taillights. The rear-window was enlarged, too.

Unchanged was the "Spitfire" eight beavering away in the engine bay. With a 3.25-inch bore and 4.88-inch stroke, it was rated at 135 bhp at 3400 rpm and 270 pound-feet of torque at 1600 revs. "It is not noted for its amazing acceleration," opined Woron.

It may not have entirely been the engine's fault. Prestomatic functioned like a four-speed semiautomatic transmission, but its selling points were smoothness and convenience, not snappy shifts. Only with some finessing of the gearbox between its high and low ranges was MT able to crack 20 seconds from 0 to 60 mph; a best quarter-mile took 22.3 seconds. Recorded fuel economy was 11.34 mpg in heavy traffic, but improved to the mid teens at steadier speeds.

Woron reported that he found the New Yorker maneuverable for its size, with "positive stopping" from the vacuum-boosted brakes.

Chrysler would become a performance machine with introduction of its Hemi V-8 in '51. However, this 1950 Chrysler New Yorker sedan had many sterling qualities. It was a big, comfortable cruiser with excellent build quality and a plush interior that satisfied many in its intended market.

# 1951 CHRYSLER WINDSOR HIGHLANDER NEWPORT HARDTOP COUPE

Nineteen Fifty-One was the year of the "hemi" at Chrysler. Saratogas, New Yorkers, and Imperials all got the brand-new 331-cid "FirePower" ohv V-8, a sensation with its efficient hemispherical combustion chambers and 180 bhp. That certainly left the Windsor series, with its staid flathead six, in the shade.

But the Windsor—now Chrysler's entry-level series with the demise of the Royal—had sufficient charms of its own. Not the least of them was its availability as a stylish Newport two-door hardtop or the chance to order it with the Highlander interior.

The pillarless hardtop body was in its second year at Chrysler, and the shape of the roof was unchanged from 1950. As was common at the time, Chrysler gave a distinct name to its hardtops: Regardless of series, they were Newports. In '51, the Windsor was offered in base and Deluxe trim, but only the Windsor Deluxe could be a Newport, which started at $2953. The Highlander interior option, which featured colorful plaid cloth on the seat cushions, back rests, and door panels, had been available on certain New Yorker and Windsor models since 1940.

Chrysler combined production figures for its 1951 and '52 cars so breakdowns by model year are not available. It's estimated that about 6400 of the 10,200 Windsor Deluxe hardtops made in the period were '51s.

Highlander
WINDSOR

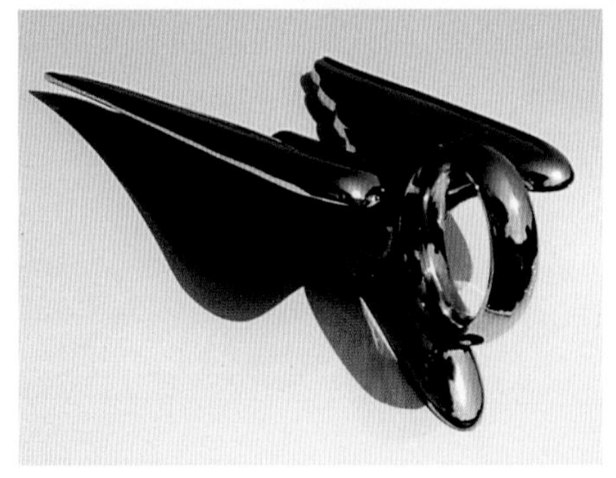

The 1951 Chryslers were extensively facelifted versions of the first all-new postwar designs the division created for 1949. Central to the new look were a two-bar grille and a laid-back, more streamlined hood than that found on the 1949-50 cars.

Chassis were carried over, too. In the case of the Windsor, that was a 125.5-inch wheelbase (139.5 for eight-passenger models) with independent coil-spring suspension in front and Hotchkiss-type drive with semielliptic leaf springs in back.

As noted, only the Windsor retained its previous engine. The 250.6-cid L-head six generated 116 bhp. It was hooked to a three-speed manual transmission with the availability of Chrysler's famous Fluid Drive system, which the featured car has.

Chrysler

# 1951 JOWETT
## JUPITER CONVERTIBLE

Jowett was an unorthodox make even by British standards. From 1910 through 1936, Jowetts were powered by horizontally opposed two-cylinder engines. A four introduced in '36 continued the "boxer" layout in which the two banks of cylinders faced each other with the crankshaft in the middle. Then, too, Jowett was located off the beaten path in the Yorkshire village of Idle. (Management must have gotten tired of "Idle factory" jokes.)

In 1947, Jowett introduced its technologically advanced Javelin sedan. The unitary body was aerodynamic for the time. Springing at both ends was by torsion bars with independent front suspension. Jowett stuck with the horizontal engine design, the like of which is only used by Subaru and Porsche today. The ohv four displaced just 1.5 liters (90 cid) and developed 50 bhp. Thanks to low weight of not much more than a ton, the Javelin had good performance for a small sedan of its time, and won its class in the 1949 Monte Carlo Rally.

JJUP 51

HB44

Encouraged by its Monte Carlo victory, Jowett decided to build a sports car using Javelin components for 1950, the Jupiter. Its backbone was a tubular chrome-moly-steel chassis. Torsion-bar suspension was retained, but Jupiter gained rack-and-pinion steering. As in the Javelin, the lightweight aluminum engine sat ahead of the front axle line with the radiator mounted behind the engine. Horsepower was increased to 60. Unusual for a sports car, the four-speed transmission retained the Javelin's column-mounted shifter.

The chassis was clothed in an aluminum convertible body with roll-up windows. The interior was well finished and roomy for a sports car—Jowett claimed three-passenger capacity. In spite of the amenities, the Jupiter weighed only 1500 pounds. Performance was good with 0-60 mph in 15 seconds, and a top speed of around 90. Performance and handling helped Jupiter win in its class at Le Mans from 1950 through 1952.

Jowett hoped Jupiter would do well on the American market, but weak crankshafts and a few other problems gave the car a reputation for unreliability. The problems were ironed out but the damage was done. Not helping was a Tom McCahill review in *Mechanix Illustrated* that said the Jupiter "dives into corners like a porpoise with heartburn and the steering is like winding an eight-day clock with a broken mainspring." *Road & Track*, which later reviewed the Jupiter, liked its handling and speculated that there might have been something wrong with McCahill's car.

Approximately 900 Jupiters had been built by the time Jowett production ended in 1954.

# 1952 HEALEY
## TICKFORD SALOON

Donald Healey is famous for the Austin-Healey, but he had an active life long before he teamed up with Austin. Healey flew for the Royal Flying Corps during World War I. Between the wars he was a successful rally driver and won the 1931 Monte Carlo Rallye. Later he was director of experimental design at Triumph. During World War II he worked on armored-car design.

After the war, Donald Healey set up a company to produce his vision of a grand touring machine. Production started in fall 1946. The new Healey featured independent front suspension and a robust frame for good handling. For power, Healey bought 2.4-liter four-cylinder engines made by Riley, another English automaker. With advanced fea-

tures such as hemispherical combustion chambers and dual camshafts mounted high in the block, 104 bhp was developed. Thanks to sound engineering and development of the basic design since 1926, the Riley engine proved to be tough and reliable.

Several small coachbuilders supplied bodies. Four-seat convertibles and coupes were the most popular, and this coupe has bodywork by Tickford. A shortened two-seat Silverstone sports car was also offered. Although the coupes had two doors and only a small back seat, Healey called them saloons—British for sedan.

Healey was quick to enter his cars in competition, and strong showings in rallies generated good publicity. In 1947, a production car was sent to Belgium for speed runs and was timed at 110.8 mph. For a time, Healey was able to advertise his car as "The fastest production car in the world."

The Healey chassis received several refinements during production. The car also gained weight but was still good for more than 100 mph. With good

handling and a willing engine, Healeys were a joy to drive. They were also expensive. An early convertible cost $7500 in the U.S. when a 1947 Ford sold for $1154.

Healey also built the Nash-Healey using Nash engines in a Healey chassis. The Nash-Healey was also expensive and Donald set about designing a more affordable car using Austin components. The Austin-Healey was such a hit at the Earls Court Motor Show that Austin offered to take over production, and work on the new car soon dominated Healey's firm.

Although the Healey was profitable, its Riley engine was set to go out of production. Healey sales ended in '54 after about 700 cars had been built.

# 1952 PACKARD PAN AMERICAN CONVERTABLE

As noted historian George Hamlin recounted in a profile of Packard Pan American designer Richard Arbib, the Pan American was one of several "sports car" ideas that the styling consultant doodled up for the Henney Body Company of Freeport, Illinois, long a supplier of Packard-based hearses, ambulances, and other professional cars.

A successful industrial designer since the late Thirties, Arbib had also worked for General Motors and, after the war, the Harley Earl Corporation. When a falling out with Earl prompted him to go freelance in 1949, Arbib contacted a previous employer, industrialist Charles Russell Feldmann. As it happened, Feldmann had just purchased Henney (for a second time) and needed help with redesigning its professional coachwork to match Packard's new 1951 "high pockets" styling. Arbib duly signed on as a Henney consultant and de facto one-man styling staff.

Feldmann wanted to expand Henney's business, in part by "doing something in the sports car area," as the designer later recalled.

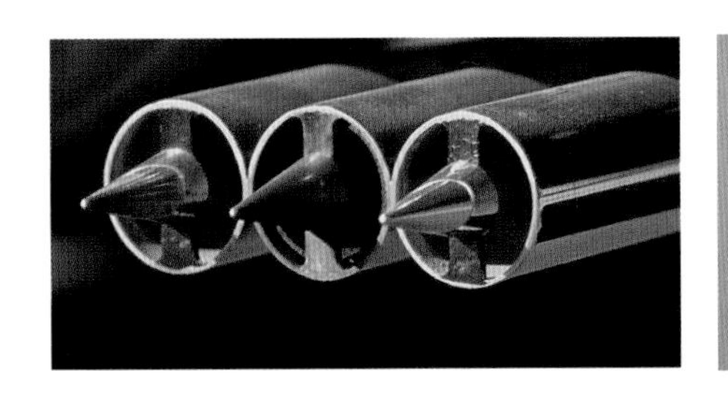

Packard Pan American

Encouraged to let his imagination roam with hopes of winning more Packard contracts, Arbib drew up a two-seat sports convertible based on John Reinhart's 1951 production design. This concept, which Arbib named the Pan American, so impressed Feldmann

that he ordered a car to be built at a reported cost of some $10,000.

It was basically a detrimmed Series 250 convertible with a rakishly sectioned body accentuated by a jaunty "continental kit," modest chrome fins above special round taillamps, a wide

hood scoop, and an extended rear deck (Arbib had removed the back seat, deeming it unsuitable for a sports car).

Headlamp bezels and the inner grille section were predictively color-matched to the body's striking gold-green paintwork. Chrome wire wheels provided an appropriate finishing touch. The interior was basically stock except for different upholstery and a repositioned steering column.

As an obvious promotional tool for both Henney and Packard, the Pan American was booked for the March 1952 International Motor Sports Show in New York, for which Arbib moonlighted as art director. The hasty scheduling left no time to fit the planned hideaway cloth top, but nobody noticed. In fact, public reaction was so positive that Packard briefly hinted at a limited run for retail sale, only to demur after ordering just five copies, which Henney supplied with tops included.

The Pan American featured here is the show car, meaning it is first of the six cars built—and the only one without its intended folding top.

# 1953 ALFA ROMEO
## 1900C GHIA COUPE

When Alfa Romeo wanted to make a splash at the 35th Turin Automobile Show, it asked six coachbuilders to build a custom body on the Alfa 1900 chassis. They were Bertone, Castagna, Boneschi, Pinin Farina, Vignale, and Turin's own Ghia.

Giacinto Ghia started building car bodies on a small scale during World War I and then prospered during the Twenties and Thirties. Ghia did its best work with sporting bodies, and Ghia coachwork graced Alfa Romeo, Lancia, and other exclusive Italian sports cars.

Ghia also built sport coupes and spiders on Fiat's reasonably priced 508 S Balilla chassis. (Balilla was named for Mussolini's Fascist youth group.)

The Fiat business expanded Ghia's output, but the factory was flattened by Allied bombing during World War II. Giacinto Ghia died soon after the war and his widow sold the firm. Carrozzeria Ghia returned to health, but at a much lower volume.

American automakers were good for Ghia in the Fifties. Chrysler Corporation discovered that Ghia could build show cars faster and cheaper than it could in-house. The Chrysler K-310, Dodge Firebomb, and DeSoto Adventurer, among other Ghia-built concept cars, adorned the Chrysler stand at Fifties' auto shows. Chrysler tapped Ghia to build Crown Imperial limousines between 1954 and

1964. Ghia also built 50 Chrysler Turbine cars in 1963—as well as the prototype. A series of Dodge Firearrow show cars led to a contract to build the Italian-American Dual-Ghia. Although popular in Hollywood (particularly with the Rat Pack), fewer than 150 were built between 1956 and 1963.

For the 1955 auto-show circuit, Ghia built Lincoln's bubble-topped Futura, which was later turned into the Batmobile. Packard's last show car, the Predictor, was constructed by Ghia for the 1956 Chicago Auto Show. Ghia didn't neglect the Italian exotic cars, building bodies for Ferrari and Maserati.

Although Ghia was successful building show cars and sports car bodies in small numbers, it didn't have facilities to build in volume. Ghia created the Volkswagen Karmann-Ghia by scaling down the Chrysler D'Elegance show car, but couldn't build in VW quantity. Karmann got the job that totaled more than 400,000 cars.

Ford acquired Ghia in 1970, and it was Ford's Italian design house for a while, as well as a new trim-level name applied to Ford Motor Company cars. The De Tomaso Pantera sold through Lincoln-Mercury dealers was a result of the Ford buyout. The Ghia name long ago disappeared from American Ford cars.

The Alfa Romeo seen here is on the 1900C Sprint chassis. The 1900 was Alfa's moderately priced postwar sedan. The 1900C Sprint was a short-wheelbase version bodied by coachbuilders and served to revive some of Alfa's prewar glamour. Only 1796 coupes were built between 1951 and '58 compared to 17,243 sedans made during 1950-55. This is the Ghia coupe built for the Turin show. At least one copy of this style was also built.

# 1953
# CADILLAC
## ELDORADO CONVERTIBLE

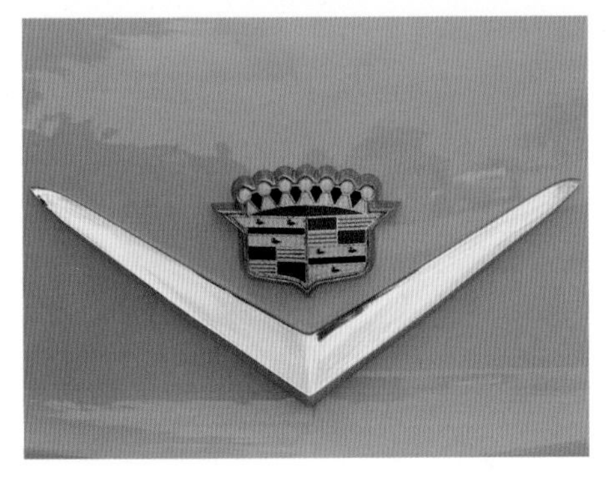

The first Cadillac Eldorado cost almost double the price of a Series 62 convertible in 1953, yet Cadillac probably lost money on every one it sold because of the extensive special tooling required for a limited-production car. The Eldorado was based on a show car displayed in 1952 to commemorate Cadillac's 50th anniversary. The name Eldorado, from

the Spanish *el dorado*, meaning "the gilded one," tied in with the marque's golden anniversary.

Because of the Korean War, there was no General Motors Motorama to show off that first Eldo in '52, but there was a Motorama tour in 1953 at which to exhibit the production Eldorado, the Buick Skylark, and the Oldsmobile Fiesta—a trio of glamorous convertibles built to raise the image of their respective marques. Eldorado's reputation got a special boost when Dwight D. Eisenhower rode in one during his presidential inauguration parade on January 20, 1953. Most presidents-elect arrived at their inaugurations in dignified limousines or phaetons. Ike, with his unassuming and sunny persona, looked natural standing up in the back of an Eldorado.

Cadillac's image was already riding high. The Sixteen helped solidify Cadillac's reputation in the Thirties. Then, starting in 1949, a powerful ohv V-8 helped establish the brand as the top American luxury car for years to come. For '53, the 331-cid V-8 was rated at 210 bhp. That was probably understated because the following year horsepower rose to 230 with no mechanical changes.

Top speed of a Series 62 coupe was 116 mph, but the 4800-pound Eldorado would have been a little slower, outweighing the coupe by 570 pounds, and coming in 300 pounds heavier than the 62 convertible. Cadillacs still rode on a prewar chassis, but that chassis did its job well, providing a smooth ride and acceptable handling for such large cars.

In the early Fifties, GM Styling under the direction of Harley Earl was in its prime, and the standard Cadillac convertible was already a dazzling design. Although the Eldorado resembled the Series 62 soft top, Cadillac had to craft a special hood, cowl, doors, and body shell for the Eldorado. Cadillac lowered the chassis an inch, and overall height was three inches lower. Mounted upon the lowered cowl was a panoramic windshield—a touch that spread to other Cadillac models in 1954. The doors had a sporting dip, and, when the top was folded, a metal cover neatly concealed it.

Chrome wire wheels were standard. Also standard were a four-speed Hydra-Matic transmission, power steering, whitewall tires, fog lights, power seat, signal-seeking radio, power windows, and windshield washers. A special leather-upholstered interior included a padded dash. The price for all this luxury was $7750, which 532 buyers considered acceptable. The Eldorado

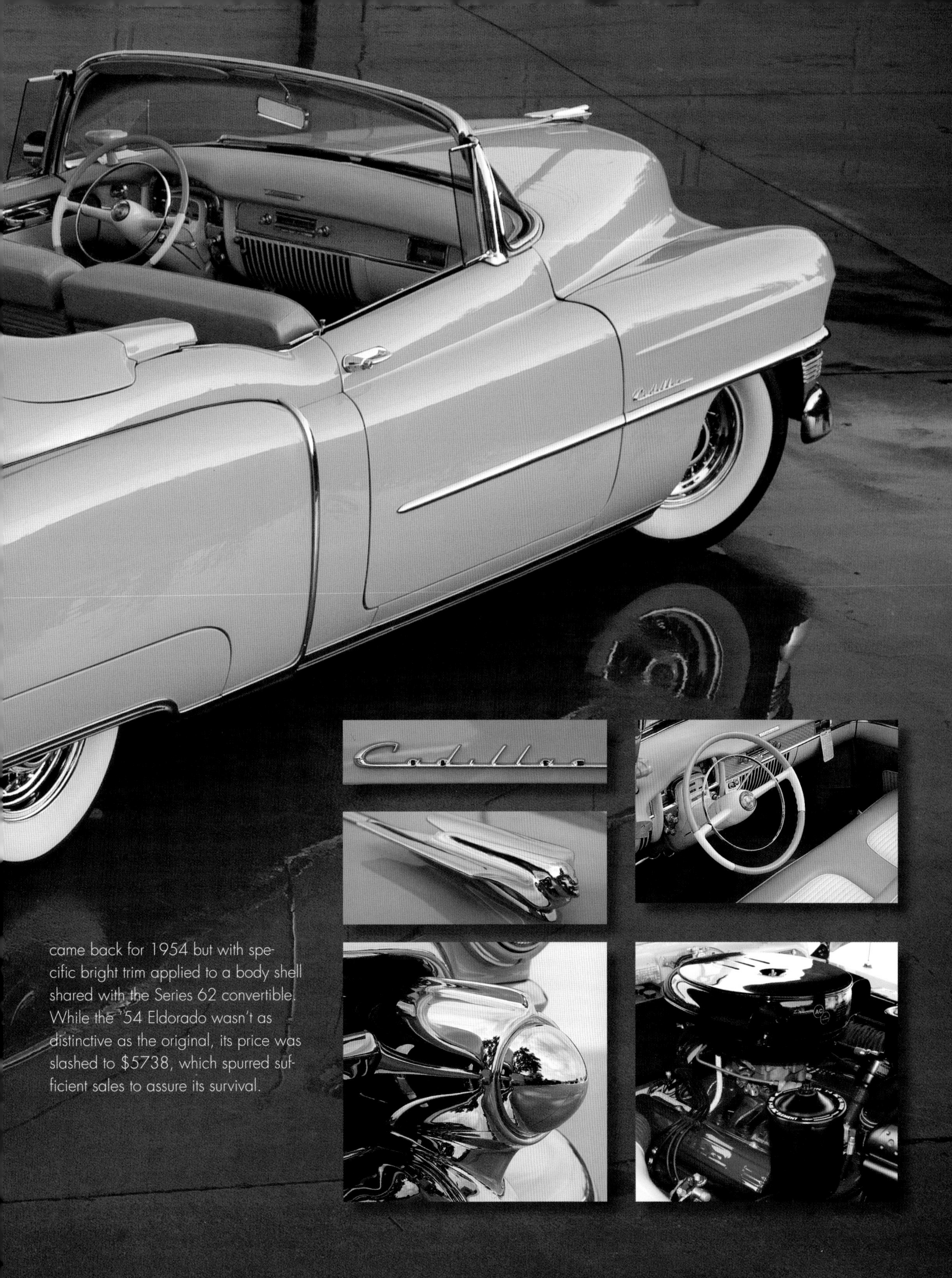


came back for 1954 but with specific bright trim applied to a body shell shared with the Series 62 convertible. While the '54 Eldorado wasn't as distinctive as the original, its price was slashed to $5738, which spurred sufficient sales to assure its survival.

# 1953 CADILLAC
## GHIA COUPES

The two 1953 Cadillacs with custom Ghia bodies featured here are surrounded in mystery. Legend says actress Rita Hayworth received one of these cars as a gift from her husband, Aly Khan. Which one? Well, nobody seems to be sure.

Fifteen years ago, *AutoWeek* tackled the story, and it, too, came up with precious little information. We've uncovered a few additional tidbits and connected a few more dots, but find ourselves with more questions than answers.

One, or both, of the cars was displayed at the 40th Salon d'Automobile et du Cycle in Paris during October 1953. The cars appear to be based on a standard 1953 Cadillac Series 62 convertible chassis. The convertible chassis was more rigid and was a better

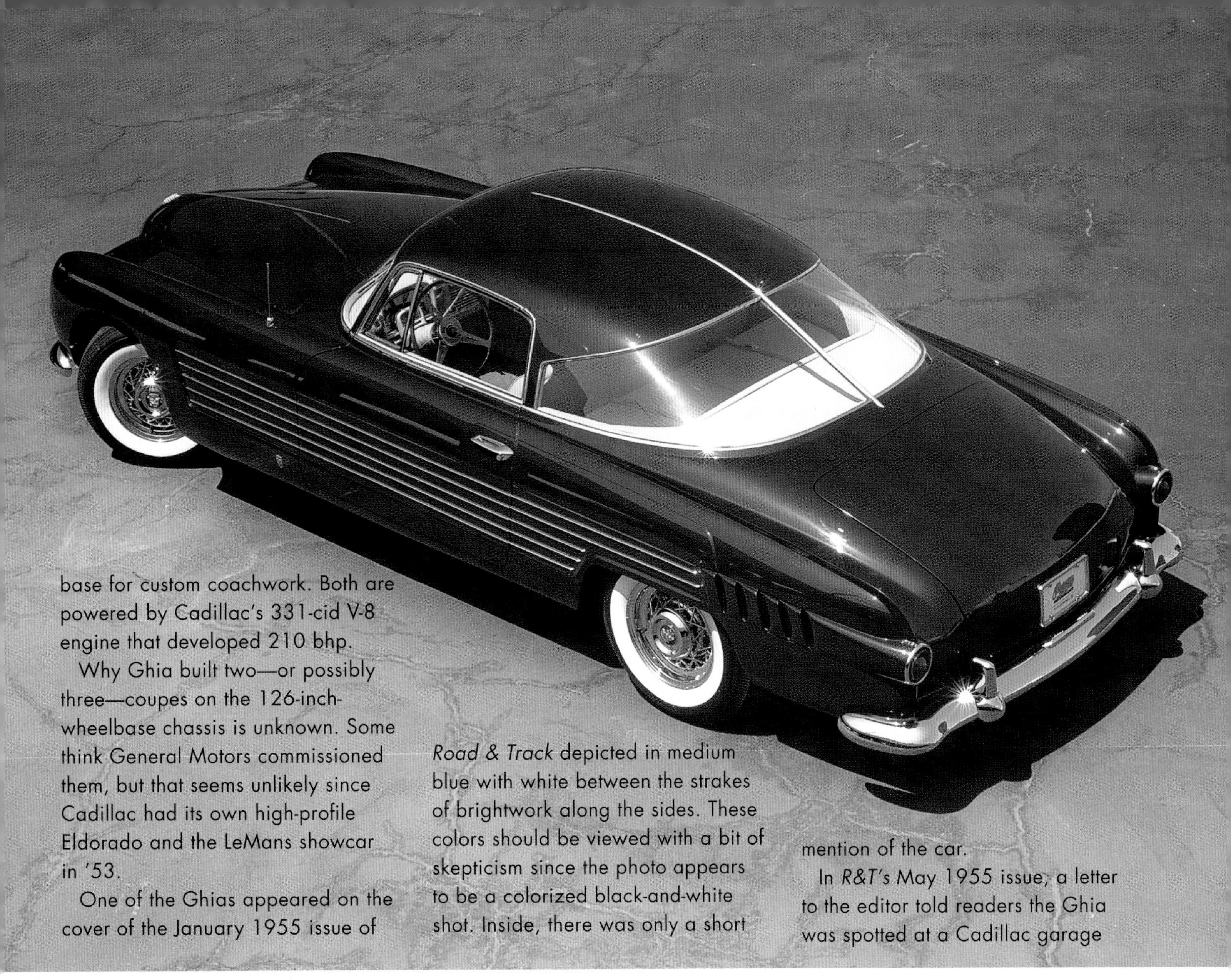

base for custom coachwork. Both are powered by Cadillac's 331-cid V-8 engine that developed 210 bhp.

Why Ghia built two—or possibly three—coupes on the 126-inch-wheelbase chassis is unknown. Some think General Motors commissioned them, but that seems unlikely since Cadillac had its own high-profile Eldorado and the LeMans showcar in '53.

One of the Ghias appeared on the cover of the January 1955 issue of *Road & Track* depicted in medium blue with white between the strakes of brightwork along the sides. These colors should be viewed with a bit of skepticism since the photo appears to be a colorized black-and-white shot. Inside, there was only a short mention of the car.

In *R&T's* May 1955 issue, a letter to the editor told readers the Ghia was spotted at a Cadillac garage

in New York City and that it was owned by John Perona. The letter also mentioned the car was for sale at $10,000.

Who is John Perona, and how does he fit into this tale? He is listed as the first owner of a Ghia-bodied 1951 Ferrari 340 America coupe, so he had previous experience with the firm from Turin. Perona also owned the legendary El Morocco nightclub in Manhattan where Rita Hayworth is said to have been a regular. Quite a coincidence, since they are both identified as the owner of a Ghia-bodied 1953 Cadillac.

Beyond this we have plenty of questions, but few solid answers. Was Perona a fan of Ghia's work who wanted something larger than his Ferrari? Did Mr. Perona and Ms. Hayworth own the same car at different times, or were they both original owners? Who owned one first?

# 1953
# KAISER
## DRAGON FOUR-DOOR SEDAN

Henry J. Kaiser was part of the consortium that completed the daunting task of building Hoover Dam more than two years ahead of schedule. No one had mass produced ships until Kaiser built World War II Liberty Ships in as little as five days. Perhaps, then, Henry could have been excused for thinking he could revolutionize car building as well. He must have soon realized that it was a bigger job than expected and that the men running Detroit were smarter than he gave them credit for being.

Kaiser was underfunded from the start and was never able to match the Big Three in engines and range of body styles. While Detroit offered ohv V-8s of ever-increasing horsepower, Kaiser had only an L-head six. The Big Three were creating new markets with hardtops and steel-bodied station wagons; Kaiser relied almost entirely on four-door sedans.

Where Kaiser did innovate was in interior design and materials. Cars of the Forties were upholstered in tan or grey cloth that didn't wear well. The average car buyer immediately applied seat covers and didn't see his factory upholstery again until trade-in time. Kaiser's color and interior chief, Carlton Spencer,

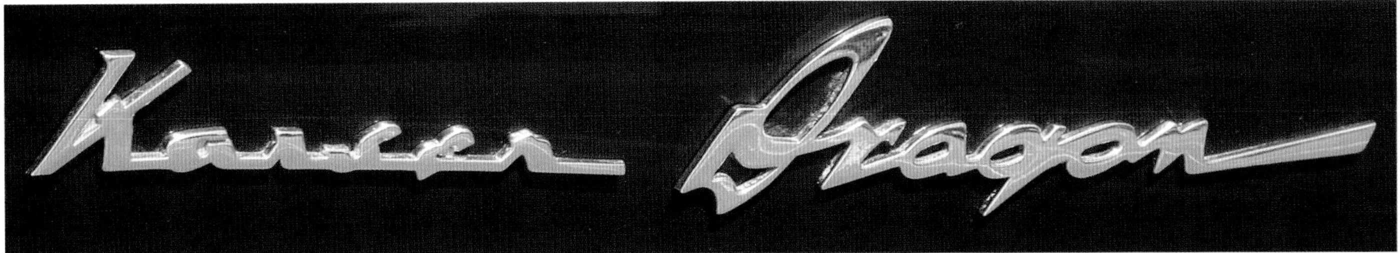
Kaiser Dragon

helped change that with durable new fabrics and vibrant colors. His masterpiece was the Kaiser Dragon.

The first Dragon appeared in 1951. The name was derived from its "Dragon Vinyl" upholstery that resembled alligator hide. For 1953, the Dragon swapped Dragon Vinyl for a combination bamboo-patterned vinyl and "Laguna" cloth—a heavy-duty Belgium linen with a very midcentury-modern pattern of overlapping rectangles. Long-filament "Calpoint" carpeting covered the floor. Dragons used more than 200 pounds of sound insulation for a quieter ride than lesser Kaisers. Dragons also had an automatic transmission, 14-carat-gold hood ornament and badging, whitewall tires, and chrome wire wheels. Although labeled a hardtop, the Dragon was actually a four-door sedan dressed up with a "Bambu" vinyl top.

There was more to Kaiser than styling. Kaiser claimed to be "America's first safety-first car" with a padded dash and large glass area. A rigid frame with a low center of gravity provided good handling and a smooth ride. However, the Dragon's list price neared $4000, well into Cadillac territory—and Kaiser's 118-bhp six was almost 100 horses down on the Caddy's V-8. That helps explain why only 1277 Dragons were built for '53.

The original owners of this Maroon Velvet Dragon received a personalized engraved medallion from the factory, which the dealer attached to the glovebox door. Although now covered with the clear-plastic seat covers that Carlton Spencer tried to discourage, the upholstery is still the original, hard-wearing Laguna cloth.

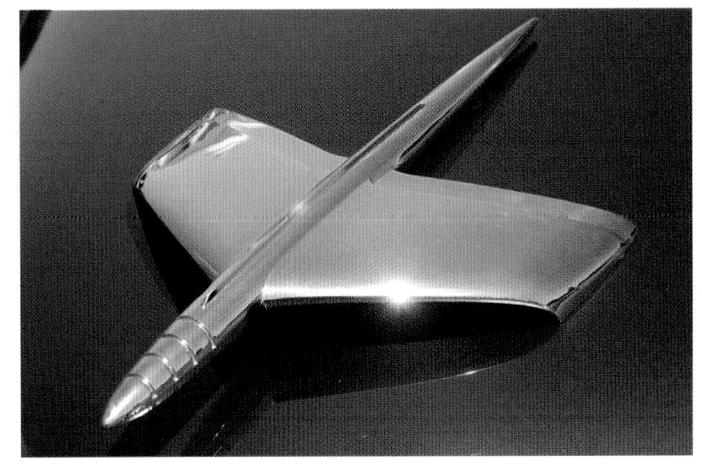

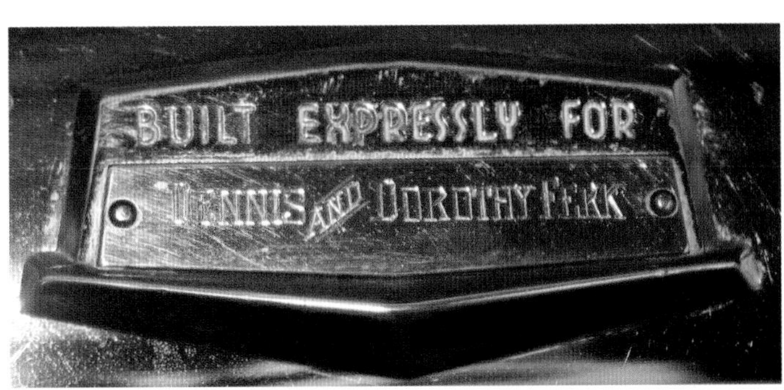
BUILT EXPRESSLY FOR
DENNIS AND DOROTHY FERK

# 1954
# ALFA ROMEO
## 1900 CSS GHIA COUPE

Before World War II, Alfa Romeo produced exotic sports cars and occupied a niche similar to that of Ferrari today. (In fact, Enzo Ferrari managed Alfa's racing team before building his own cars.) But to survive in the postwar European market, Alfa Romeo decided to market a mass-produced sedan.

The 1900, introduced in 1950, broke with Alfa traditions and opened new markets for the Milan automaker. The prewar ladder frame was replaced by unibody construction and an expensive independent rear suspension was replaced by a cheap, but well-located, solid axle. Coil springs were used on all four corners. European racing convention favored right-hand drive, and race-oriented Alfa Romeo always put the steering wheel on the right—until the 1900. The new engine was a four instead of a six or eight, but retained dual overhead cams. It was a jewel of an engine with good power for its size and a wonderful exhaust note. The 1900 proved a practical, reliable car, but still carried on the Alfa tradition of performance and good handling. Alfa sloganeering called it "the family car that wins races."

Alfa hadn't totally forgotten its storied past, though. The 1900 was a joy to drive and Alfa knew it could make it into something more exciting. In 1951, a short-wheelbase 1900 was made available for custom work. Italian coachbuilders used it for some of their most unusual efforts. The Disco Volante—"flying saucer"—resembled its namesake, as did the tailfinned BAT. More successful was a

low-production series of attractive 1900 coupes and convertibles. Lighter weight and dual carbs increased performance. For '53, an increase in engine size and a five-speed gearbox (in place of a four-speed) hiked performance further. Top speed for coupes was more than 110 mph and 0-60 came in at around 11 seconds. Fuel consumption was a thrifty 22 mpg.

The success of the 1900 line allowed Alfa Romeo to develop the smaller and even more successful Giulietta series that established Alfa in the U.S.

Our featured car was displayed at the 1954 Los Angeles Auto Show, where it was voted most beautiful car. The one-off body by Ghia bears a strong resemblance to the Ghia-bodied Ferrari 195 Berlinettas. The car cost almost $20,000 when a regular 1900 coupe cost only $6100.

The car's first owner was Al Williams, proprietor of the El Pappagallo restaurant in San Francisco's Fairmont Hotel. He had a custom back seat added for his dog.

# 1954 CHRYSLER
## GS-1 SPECIAL COUPE

Practically from the beginning of human experience it's been true: People often intensely want what they absolutely can't have. Take cars, for instance. Dynamic "dream car" designs often excite showgoers with the promise of styling and features that then run up against the cold, hard realities of manufacturing capability and production economics, stopping them dead in their tracks.

Once in a while, though, an inspiring design breaks free of this bottom-line tyranny—or, at least, very nearly does. The 1954 Chrysler GS-1 Special, our featured car, was just such a vehicle. For all "practical" purposes, the GS Special was a Virgil Exner show-car design modified—though not unrecognizably—into a limited-edition grand tourer.

Soon after Exner joined Chrysler Corporation, he began a series of dream car designs, elements of which would eventually work their way onto production cars. These provocative specials were handcrafted in the Turin, Italy, shops of Carrozzeria Ghia. Among them, the 1952 Chrysler Special featured bladelike fenders front and rear; round wheel openings; a low, light "greenhouse"; and headlamps that projected from either side of a trapezoidal grille. A modified version of this Paris Auto Show car was built for C.B. Thomas, president of Chrysler's export operations, in 1953, and this so-called "Thomas Special" gave rise to thoughts of a limited-production car. The essential design would be retained, and Ghia would build the cars.

Chrysler

The GS Specials were built on a 125.5-inch-wheelbase New Yorker chassis. A Chrysler 331-cid "hemi-head" V-8 with 195 horsepower, hooked to the new PowerFlite two-speed automatic transmission, made up the powertrain. The 1954 Chrysler parts bin was also tapped for items like bumpers, taillights, the steering wheel, and major gauges. Chrysler accessory wire wheels were also used on the cars. This car had a back seat, while the Chrysler Special's rear compartment was filled by a spare tire and 40-gallon fuel tank. The GS-1s were sold exclusively in Europe by the corporation's French distributor.

A production run of 400 has often been cited for the GS-1. But modern research suggests that perhaps only a dozen Thomas Specials and nine GS-1s were ever actually made.

# 1955 FORD
## THUNDERBIRD CONVERTIBLE

Style, power, and comfort: Put them together in an automobile and it's sure to be a winner. That's the combination that worked so well for the Ford Thunderbird when it was introduced for 1955.

The introductory T-Bird had style in spades, owing to its low, two-seat bodywork. (Ford pointed out in advertising that the Thunderbird's height from the cowl to the ground was just slightly more than three feet.) It had a 102-inch wheelbase—more than a foot shorter than a '55 Ford sedan—yet borrowed heavily from the many new styling touches adopted for the year's full-size Fords. The 'Bird managed to avoid being cluttered with excessive chrome trim or outlandish two-tone paint schemes.

The power part of the equation came from the 292-cid version of the corporate "Y-block" ohv V-8 shared with Mercury. Equipped with a Holley four-barrel carburetor and dual exhausts, it made 193 bhp at 4400 rpm in cars built with the standard three-speed manual transmission, or 198 bhp in those fitted with the extra-cost Fordomatic automatic gearbox. Overdrive was also available for stickshift cars.

Though it was inspired by the many mostly foreign-built sports cars that were capturing the imagination of American enthusiasts in the early Fifties, the Thunderbird was built like a little luxury car. Its family car-like suspension delivered a softer ride than "pure" sports cars (and would be made softer still in '56). Roll-up windows and the availability of a removable fiberglass top lent more all-weather comfort than the folding tops and snap-in curtains associated with most other two-seaters. A standard telescoping steering column helped drivers find an optimal position behind the wheel. Optional power assists cradled passengers in the lap of luxury.

While other sports cars for sale in the U.S. sold in small numbers, the first-year T-Bird managed to win over 16,155 customers. Though great numbers for

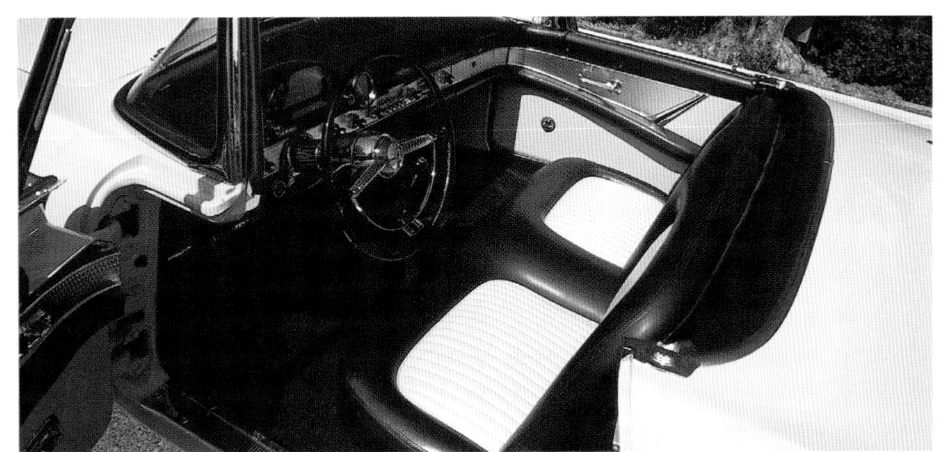

the market class, they were small by Ford Motor Company standards; come 1958, a far more popular four-seat Thunderbird replaced the winsome two-place cars of 1955-57.

This '55 Thunderbird is painted Goldenrod Yellow—one of five colors available for that year's T-Birds. It is equipped with automatic transmission, power windows and seat, a radio, and a heater.

# 1955 LINCOLN
## CAPRI CONVERTIBLE COUPE

The 1955 Lincolns were about as new as they could be, considering they came in body designs that already had three seasons on them. Aside from an obligatory touch-up of external details, the '55 Lincolns sported an all-new powertrain.

By boring out its ohv V-8 engine, Lincoln increased displacement from 317.5 cubic inches to 341. The extra cubes, a new high-lift camshaft, and standard dual exhausts produced almost 10 percent more power than in '54. There was also a half-point compres-

sion boost to 8.5:1 that spurred Lincoln to refer to its engine as having "new super-compression power". Thus, the "Fleet-Power" mill now made 225 bhp at 4400 rpm and 332 pound-feet of torque at 2500 rpm. In place of the Hydra-Matic transmission Lincoln had been buying from General Motors, it now had its first homegrown automatic: Turbo-Drive. It consisted of a torque converter and three planetary gears, and offered the availability of low-gear starts if the accelerator was floored. The new powertrain propelled the two-ton-plus 1955 Lincoln to 60 mph in 12 seconds.

Frontal styling was freshened with a new grille made up of horizontal bars. Headlamps were newly hooded in period style. At the rear, the impression of length was enhanced via rear-leaning fender ends. Bodyside sculpting, which

WISCONSIN
ROCKN 55
FEB

had risen high on the rear quarters of 1952-54 Lincolns, was now confined to the lower half of the body and pointed forcefully forward.

There were parts of the existing Lincoln package worth keeping. The 1955 models continued on the same 123-inch wheelbase as before, and used the same ball-joint independent front suspension that won the "Road Race Lincolns" (so-named because of their prowess in Mexico's Carrera Panamericana) raves for their handling.

In spite of the improvements, Lincoln suffered saleswise in 1955, which was a phenomenal year for the auto industry as a whole. Model-year production slumped to 27,222, a 26 percent drop from the year before. (Even the '52s, built under government-imposed production limits due to the Korean War, were more numerous—by 49 cars.)

Just 1487 of the '55s were Capri convertibles. Capri was the top trim level and the convertible was only offered in the Capri line. Accessories on this car include road lamps mounted in front-bumper pods, dual spotlights, door-handle shields, and a locking fuel-filler cap. The body is sprayed in Cashmere Coral; the interior—which features leather upholstery—is in Cashmere Coral and Ermine White.

# 1956 DESOTO
## FIREDOME SEVILLE HARDTOP SEDAN

Chrysler products of the early Fifties were finely engineered, but unfortunately the cars looked like they had been styled by engineers, too. The visual part of the equation—likely the most important piece on the showroom floor—was addressed for 1955. The corporation's cars evolved from frumpy to Fifties fabulous thanks to Director of Styling Virgil Exner's "Forward Look."

DeSoto benefited as much as any of the five brands from this process.

Though its overall size hadn't changed much from 1954, appearance was dramatically different and the cars looked longer, lower, and wider. The design extended to careful selection of interior and exterior colors as well.

This was the era of the expected annual styling change, and for '56, DeSoto didn't disappoint. Up front, the marque's traditional toothy grille was replaced with a mesh unit suspended from vertical towers that looked much like bumper guards but actually housed the turn signals.

The big news was out back. As he did to all of Chrysler's 1956 offerings, Exner added tailfins. The fins began near the rear window and gracefully

rose in a straight line as they extended to the end of the fender. Viewed from the rear, the peaked fin formed the top of a forward-leaning bright-metal panel that housed three vertically stacked circular taillamp lenses. DeSoto marketers called this arrangement the "Control Tower." Meanwhile, a new rear bumper blended smoothly with the reworked fenders.

The fins transformed the DeSoto's rear end, adding a sense of drama where the original rear fender and taillamp treatment had been somewhat generic in comparison. The color-sweep areas on the bodysides were revised at the rear as well, with the upper molding helping to define the lower edge of the fin.

Other talking points were minor. They included a 12-volt electrical system, improved brakes, gas-fired heater, and an optional Highway Hi-Fi record player.

The DeSoto lineup expanded for '56 thanks to the addition of a four-door

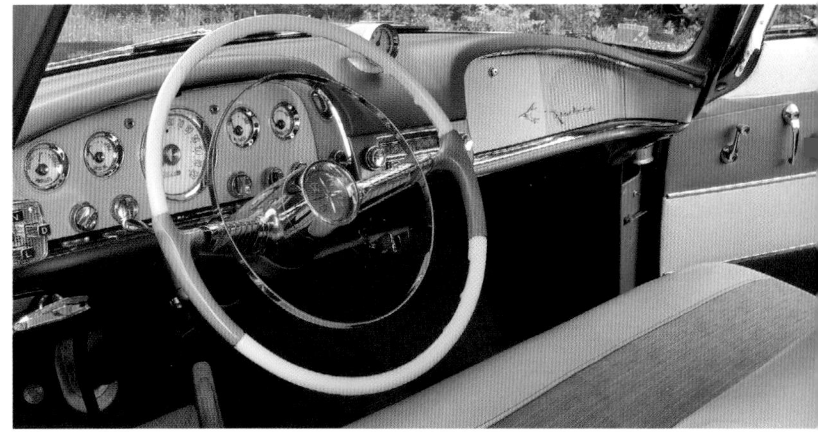

hardtop body style that was available in both DeSoto series. There were two Firedome four-door hardtops—the $2833 Seville and a better-trimmed Sportsman that listed for $120 more. Each came standard with a 230-bhp version of DeSoto's 330-cid "hemi-head" V-8. In the top-of-the-line Fireflite series, the new four-door hardtop was only sold as a Sportsman. It priced from $3431. Its version of the 330 V-8 had a four-barrel carburetor and made 255 horsepower.

Of the three new four-door hardtops, the lowest-priced Firedome Seville proved most popular, with deliveries totaling 4030. The Fireflite Sportsman rang up 3350 sales, and the Firedome Sportsman another 1645.

Our featured car is a 1956 Firedome Seville four-door hardtop that except for an exterior repaint largely remains in original condition.

OLIO
TEMP. ACQUA
BENZINA
km/h

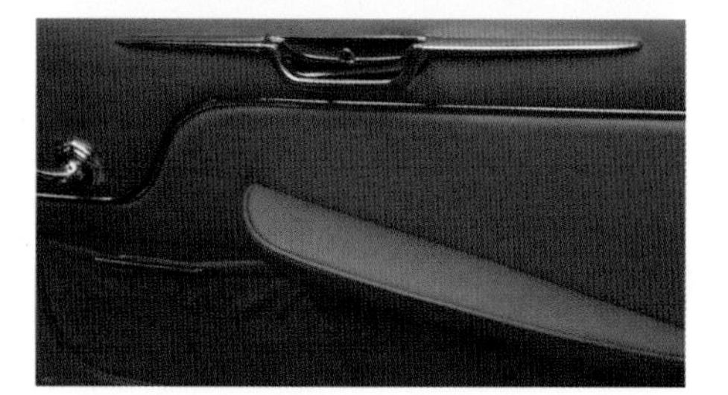

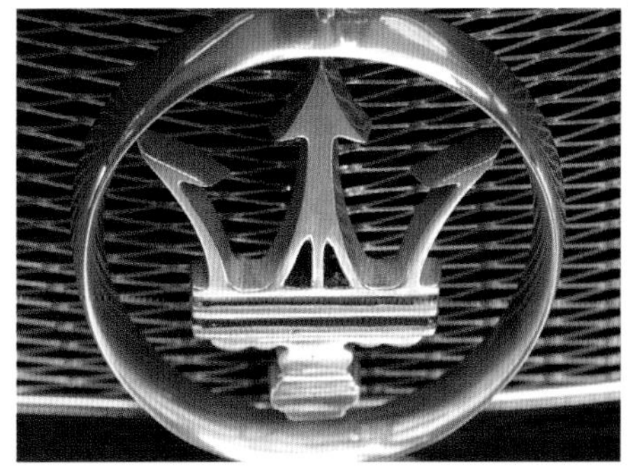

# 1956 MASERATI
## A6G 2000GT COUPE

While the Maserati nameplate has been part of the automotive world since the Twenties, the five car-building Maserati brothers initially busied themselves solely with competition machines. It wasn't until after World War II that the marque's trident logo appeared on roadgoing grand tourers.

By the mid Fifties, Maserati was starting to hit its stride, as evidenced by the A6G 2000GT coupe that is seen here. The 2000 was the beneficiary of a double-overhead-camshaft six-cylinder engine that greatly enhanced performance. Meanwhile, a number of Italy's leading coachbuilders contributed lovely body designs. Among them was the shop of Allemano from Turin, which fashioned the body of the featured car.

The Maserati on these pages wasn't actually the handiwork of any of the Maserati brothers. In 1938, they sold out to industrialist Adolfo Orsi, who moved the operation from the Maseratis' native Bologna to Modena. When their service contract with Orsi expired at the end of 1947, the three surviving brothers moved on to form OSCA—the Maserati name remained the property of their former employer.

Shortly before their departure, though, Ernesto Maserati had worked up a roadable 1.5-liter ohc six-cylinder sports car,

435
FDY

Maserati 2000 G.T.

MASERATI

the A6. The Orsi-controlled Maserati works put 65-bhp two-seat A6 coupes and convertibles on sale in 1948. Three years later, a 2.0-liter ohc six was fitted into a second-generation car known as the A6G. It made 100 horsepower. Produced by the handful into 1954, the A6G enjoyed the attentions of coachbuilders Frua, Farina, Vignale, Ghia, and Bertone.

It was at the 1954 Paris Salon that Maserati first showed the next evolution of the A6, the A6G 2000. Its distinguishing technical feature was an engine that, while still a 2.0-liter and fed by triple Weber carburetors, was completely different from its predecessor. For starters, it was a dohc design. Where the prior model's cylinder dimensions were undersquare, the 2000's were oversquare (and displacement was 31cc greater). With 150 bhp at 6000 rpm, it moved this latest A6G to 60 mph in 10 seconds; top speed was said to be about 130 mph.

Like earlier A6s, the 2000 utilized a four-speed transmission, recirculating-ball steering, and hydraulic drum brakes all around. Independent coil-spring suspension in front and a solid rear axle with leaf springs were continued, as was the 100.4-inch wheelbase, but wider 6.00×16 tires now wrapped around the Borrani wire wheels. Body-on-tubular-frame construction was another carryover feature.

Regardless of which carrozzeria handled the styling, the 2000 had a lower, wider interpretation of the divided Maserati grille. Just 60 examples of the A6G 2000 were built through 1957. The featured car is one of the 21 Allemano-bodied coupes produced during that run.

1956
OLDSMOBILE
SUPER 88 HOLIDAY
HARDTOP SEDAN
HOLIDAY

It's hard to believe that the car on these pages is original from the tires up. This 1956 Oldsmobile Super 88 Holiday four-door hardtop was purchased by an elderly lady in Kentucky and used by her housekeeper for shopping. A nephew inherited the car, but

didn't drive it—although he did start the engine occasionally before selling it.

Low mileage is one of the secrets of the Olds's eternally young look: There are only 6700 miles on the car, which has required no mechanical or cosmetic work. Even the tires—again original—

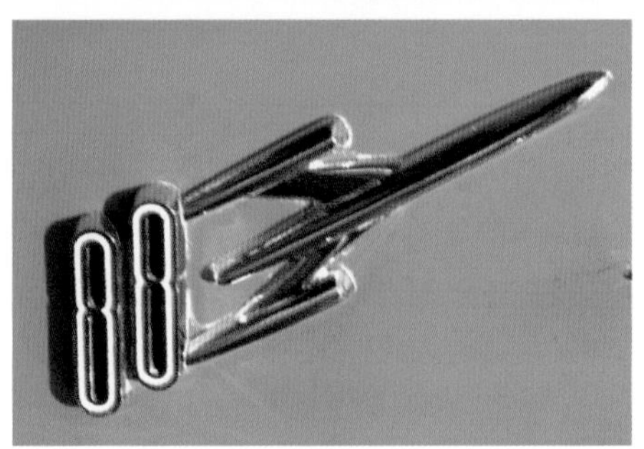

are in surprisingly good condition.

When new, there was no need to baby the Olds Super 88. *Science and Mechanics* magazine clocked 0-60 mph in 10.2 seconds and stated that the Olds "clobbers the 11.4+ seconds averaged by the seven 1956 test cars we have tested so far this year." In a Q & A format for its '56 buyer's guide, *Motor Trend* praised it this way: "Q. Does it [Oldsmobile] have good performance? A. Almost a superfluous question—it's an Olds, and the already-hot Rocket engine has been boosted in horsepower. A real delight for the would-be dragster. . . ."

The boost in horsepower was the result of an increase in the compression ratio from 1955's 8.5:1 to 9.5:1 for '56. The horsepower rating of Oldsmobile's ohv Rocket V-8 engine used in the Super 88 jumped from 202 to 240. Top speed was estimated to be 110 to 113 mph. The Rocket V-8 of our featured car is mated to a new-for-'56 Jetaway Hydra-Matic Drive automatic transmission. For $15 more than Olds's base Hydra-Matic, Jetaway added

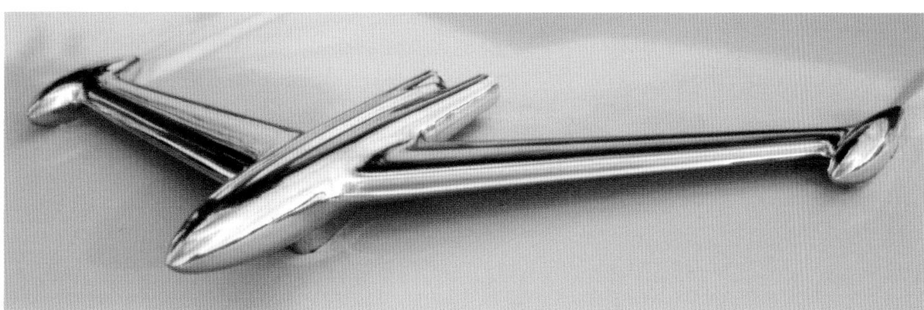

a second fluid coupling for smoother operation as well as a "Park" position in the quadrant that eliminated the need to apply the parking brake.

Just as Olds had massaged the powertrain for '56, it also revised styling with a new concave grille. The Holiday two-door hardtop had been joined by a four-door Holiday in '55. By '56, the hardtop sedan was the most popular body style in the Super 88 range with 61,192 sold at $2881 to start.

The Super 88 was in the middle of Oldsmobile's three-series lineup. The base 88 and Super 88 shared a 122-inch wheelbase. The Super boasted better trim and 10 more horsepower. The Ninety-Eight rode a 126-inch wheelbase and had standard power steering and Jetaway transmission. The Super 88 had the best power-to-weight ratio of the three, which would have allowed this car's first driver to make quick work of those shopping trips.

# 1956 PACKARD
## CLIPPER CUSTOM FOUR-DOOR SEDAN

Packard's hard-driving president, James Nance, was determined to reestablish Packard in the luxury field. To do that he planned a long-overdue separation of Packard's luxury and medium-priced lines. A new model name was needed for the lower-price line, and the Clipper name from the Forties was resurrected for the job.

While the company worked feverishly to challenge Cadillac, some might think that Clipper models were neglected. That wasn't the case because the engineering and styling revolutions of the senior Packards were shared by Clippers.

Packard's smooth, reliable, straight eights were the luxury-car standard in the Twenties and Thirties, but Cadillac's high-compression ohv V-8 set the pace in the postwar world. For '55, Packard brought out its own ohv V-8 and beat Cadillac in the horsepower race. Clippers used a smaller version of the V-8. Still, with 352 cid and 275 bhp, the '56 Clipper Custom had the biggest, most powerful engine in its price range.

Perhaps Packard's most successful engineering advance was the self-leveling Torsion-Level front and rear torsion-bar suspension. Packard not only had the best ride but was considered by many to be the best-handling American car of its time. To demonstrate the new suspension, a Lincoln, a Cadillac, and a new Packard were driven over a rough railroad crossing. The Lincoln was damaged, the Cadillac bounced, but the Packard took it in stride.

Styling was perked up with a wraparound windshield, new grille, and—on '56 Clippers—a broad band for two-tone color running along the body sides and "boomerang" taillights that became a favorite with customizers.

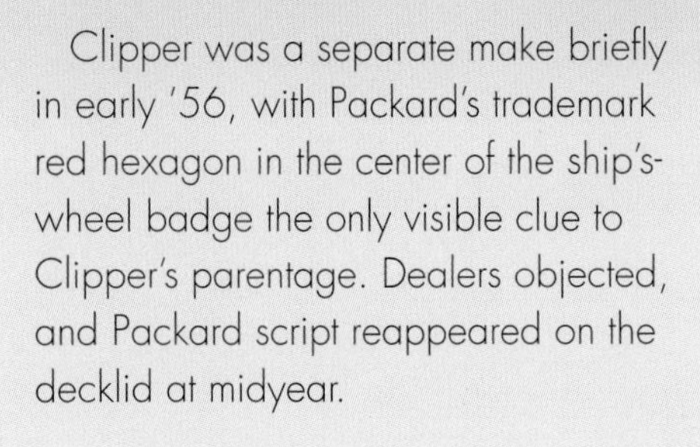

Clipper was a separate make briefly in early '56, with Packard's trademark red hexagon in the center of the ship's-wheel badge the only visible clue to Clipper's parentage. Dealers objected, and Packard script reappeared on the decklid at midyear.

The 1955-56 Packards and Clippers were great cars that should have saved the company, but the new cars were built in a new factory. Both factory and cars had teething problems that led to severe delivery shortages and quality problems among the 1955 models.

The 1956 models were a fine package, but sales were worse than ever, and quality-control problems continued. For 1957, once-proud Packard was reduced to being a retrimmed Studebaker President with '56 Clipper taillights. After 1958 Packard was gone.

# 1956 STUDEBAKER
## GOLDEN HAWK HARDTOP COUPE

Studebaker was struggling mightily in the mid Fifties, but if it had one thing going for it—apart from an infusion of Packard's money, which no longer was too plentiful at that—it was a particularly beautiful body for two-door models. That was the so-called "Loewy coupe" design done by Raymond Loewy employee Bob Bourke for the all-new line of 1953 Studebakers.

In 1954, Studebaker and Packard were merged, and the new S-P management decided it needed someone less expensive than the renowned Loewy to style its cars in the future. But while it turned over the styling of its 1956 sedans and station wagons to Vince Gardner, it gave Loewy's firm one last crack at facelifting the stunning coupes and hardtops it had created. The result was a series of four cars in a new family of Hawks. The most prestigious of them all was the befinned Golden Hawk.

In place of the drooping hood and low grille of the 1953-55 Studes, Bourke substituted a taller hood that tapered back from a trapezoidal central grille. Horizontal slots flanking the grille bore hints of the previous design, though. In back, the rounded decklid of

the past was updated with a flat, ribbed section to which further attention was called via the use of contrast-color paint on two-tone cars. Low-line Flight and Power Hawks utilized the pillared roofline of the 1953-55 Starlight coupes, but lusher Sky and Golden Hawks adopted the related Starliner pillarless hardtop roof.

The Golden Hawk's emblems of rank began, predictably enough for the Fifties, with tailfins. Vertical fiberglass blades were grafted atop the rear quarter panels. (Though fairly modest for the era, Bourke didn't care for the Golden Hawk's fins.) Wheel-arch moldings linked by ribbed lower-body trim also distinguished the Golden Hawk, as did a wide bright band at the rear of the roof, a touch carried over from the '55 President Speedster.

While lesser Hawks were powered by Studebaker-built six-cylinder and V-8 engines, the Golden Hawk received Packard's 352-cid V-8. With 275 bhp on tap, the engine could propel the Golden Hawk to top speeds of more than 120 mph, but its weight made the car somewhat nose-heavy and compromised handling. A three-speed manual transmission was standard, with overdrive and Packard's Twin-Ultramatic automatic available.

The power and panache sold for $3061 without any options. Golden Hawk orders totaled 4071 for 1956.

This '56 Golden Hawk is packed with desirable features like the automatic transmission, heater, power brakes, radio with twin rear antennas, and wire wheel covers.

# 1957 BUICK

## CENTURY CONVERTIBLE COUPE

Twenty-one years after its introduction, the Buick Century was still playing its historic role as a "banker's hot rod." The Century series bowed in '36 when Buick dropped its biggest engine in a smaller chassis.

Thanks to the resulting power-to-weight ratio, the Century name was no idle boast: By its second season, the car could do 100 mph. The same formula still worked in 1957, when the Century

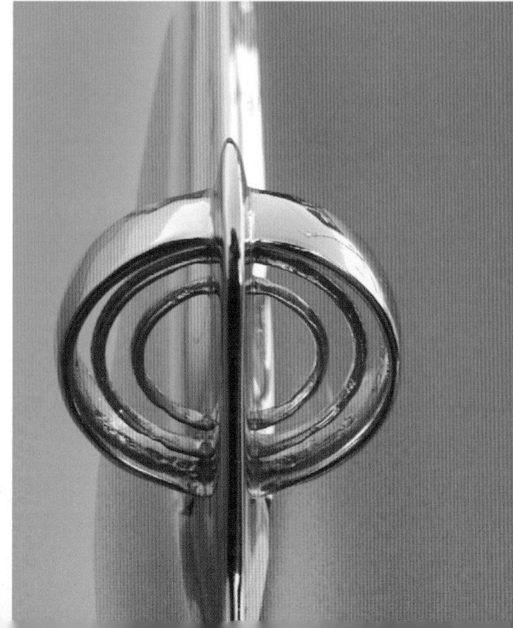

1957

BUICK

CALIFORNIA
BRR 303

shared its 122-inch wheelbase with the entry-level Special but had the more powerful engine of the larger Super and Roadmaster series. *Motor Life* estimated a top speed well beyond the century mark at 125 mph.

All '57 Buicks had a 364-cid ohv V-8. The Century and bigger Buicks had a four-barrel carburetor and 10.0:1 compression ratio that produced 300 bhp—a 50-horsepower advantage over the Special. By comparison, the standard Cadillac engine of that year produced the same 300 bhp and was only one cubic inch bigger.

To make the most of that power, Buick had recently improved its smooth, but formerly sluggish, Dynaflow automatic transmission for faster acceleration. With this powerteam, *Motor Trend* timed the 0-60-mph sprint in a '57 Century at 10.1 seconds. Meanwhile, handling was better thanks to a lower ride height and Buick's first ball-joint front suspension—though "better" is a relative term as this was a Fifties Buick with a suspension tuned more for a soft boulevard ride than sports car handling.

Buick had new bodies for '57 that were longer and lower, yet carried over the styling themes of recent years to maintain continuity. Just in case the changes were so subtle that one didn't recognize a new Buick, the badge on the center of the grille proudly proclaimed "1957."

Making use of the General Motors B-body shell, the 4085 Century convertibles built for 1957 had a curb weight of 4234 pounds and priced from $3598 in standard trim.

The Century convertible featured on these pages includes an optional Sonomatic radio that the driver was able to tune using a floor-mounted pedal to seek the next station. Other extras on this "banker's hot rod" include power assist for brakes, steering, front seat, windows, and radio antenna.

# 1957 MERCEDES-BENZ
## 220S CONVERTIBLE COUPE

This 1957 Mercedes-Benz 220S cabriolet looks much different from when it was discovered on an M-B dealer's lot in 1980 with a badly rusted trunk section, the wrong engine, and its suspension control arm embedded in the asphalt. Now it's an attractive, fun-to-drive parade car.

Like other automakers after World War II, Mercedes's first postwar car was a prewar model. It wasn't until 1951 that M-B brought out all-new cars. The 300 series was the prestige range with a flagship sedan and, later, the high-performance 300SL. The smaller, less expensive 220 was closely related to the 300. Each was powered by an ohc six from a successful engine family that would be in production through 1972. The 220 used a smaller 2.2-liter version. It was mated to four-speed manual transmission with column shift. The chassis was advanced for the Fifties with full independent suspension; the rear suspension utilized swing axles. However, the 220 retained separate

20 40 60 80 100 120 140 160 180

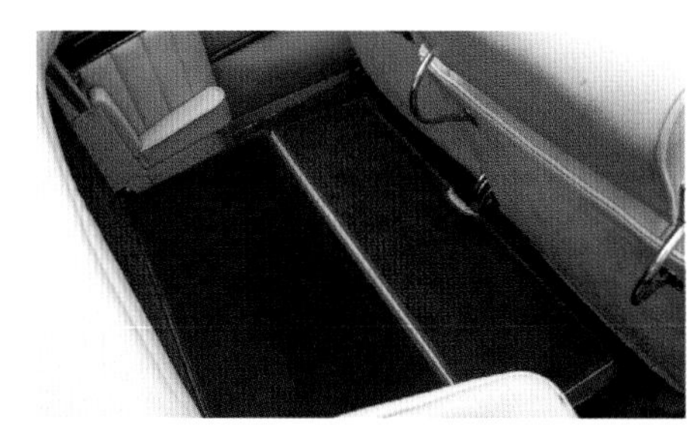

220 S
Chula Vista
NOV
CALIFORNIA
DNAGER 3
DNAGER CORPORATION
D

body-on-frame construction until the envelope-body 220a sedan introduced in spring 1954 shifted to a unitized platform. Convertible bodies were still made the prewar way with steel panels over a wood frame.

During 1956 the 220 gave way to the more modern 220S. All body types now came with unit-body construction. Coupe and convertible styling looked more modern with the separate, prewar-style fenders and running boards replaced by an up-to-date body with integrated fenders. Only the sedan was offered initially but the coupe and convertible joined the range later in '56. The two-door models rode a 106.3-inch wheelbase that was 4.7 inches shorter than the sedan. For the 220S, the six gained dual carburetors and produced 112 bhp. Top speed was around 100 mph and 0-60 mph was achieved in 15-17 seconds.

Although not a hot rod, the 220S was one of the more elegant convertibles of the late Fifties. The interior was covered in leather and wood veneer, and this well-crafted luxury had a price. The 220S cabriolet cost $7138—about the same as a Cadillac Eldorado Biarritz convertible in '57.

In 1957, Mercedes hoped to expand its sales in the United States. To support this, parent company Daimler-Benz signed a distribution agreement with Studebaker-Packard Corporation. The current-day American branch, Mercedes-

Benz USA, wasn't organized for another eight years.

This 220S convertible looks like a prizewinner but, it's not displayed at car shows. Instead, the cabriolet is a regular participant in parades. The back seat of the Mercedes is cramped to sit in, but it's sized just right for kids to stand up in during parades.

It is a European-market car with a back seat that folds down to hold luggage, and the owner has a set of period-correct factory luggage to fit. An unusual feature for a Fifties car are the detachable headrests.

1958
FERRARI
250 GT ELLENA
COUPE

Ferrari has long been recognized as a top builder of exotic sports cars. That tradition can be traced back to the 250 series of the early Fifties, when Ferrari got serious about building production (by exotic-car standards) sports cars.

Company founder Enzo Ferrari managed the Alfa Romeo racing team before World War II. Starting in 1947, he built race cars and a few road-going sports cars in small numbers. The 250 sports-car series began with the 250 Europa GT of 1954-55, followed by the 250 GT coupe that was introduced in 1956.

The GT coupe had a Pinin Farina-styled body, but Pinin Farina didn't yet have the production capacity to build in sufficient numbers, so Carrozzeria Boano built around 80 bodies. When Mario Boano left his company in 1957 to work in Fiat's styling office, his son-in-law Ezio Ellena took over operations and changed the name to Carrozzeria Ellena.

Along with the name change, the cars changed as well. They had a slightly higher roof and did away with door vent windows. About 50 Ellena coupes were built before production ended in 1958 and a new Pinin Farina -designed and -built coupe replaced

the Ellena. The 250 GT Ellena coupe is similar to both a GT-racer version badged 250 GT Berlinetta (also known as the "Tour de France") and to the highly prized 250 GT Spyder California convertible.

The simple, elegant coupe body rode on a rigid tubular frame with a 102-inch wheelbase and weighed about 3000 pounds. *Sportscar Quarterly* said of the 250 GT's handling, "Roadholding is typical of a race-bred car: just perfect. No roll on S-bends, no appreciable heeling even on tight corners." The Ferrari V-12 displaced 3.0 liters; the 250 name came from the 250cc displacement of each cylinder. The engine had one overhead camshaft per bank and three two-barrel Weber carburetors. Ferrari rated the engine at 240 bhp, but actual horsepower was 220-230. Each engine was tested on a dynamometer. Ferrari installed the engine when the chassis returned from the coachbuilder. A racecar driver or favored customer would get a 230-horsepower engine, while "average-Joe" Ferrari buyers got 220-horse units. Whatever the rating, the 250 GT was quick, with 0-60 mph in 5.9 seconds and a top speed of 127

to 157mph with available axle ratios ranging from 3.66:1 to 4.85:1. All had a synchromesh four-speed manual transmission.

*Car and Driver* said of the Ferrari's performance, "The combination of this kind of low-speed pulling power with screaming, slam in the back acceleration that hustles you from 0 to 120 mph in 24 seconds is just one of the unusual virtues of this rather unique machine". Stopping the 250 GT from those speeds were large 14-inch aluminum-drum brakes mounted behind 16-inch wire wheels.

This car is the fifth from the last of the Ellena coupes built.

# 1959 CHRYSLER
## NEW YORKER HARDTOP COUPE

The smartly elegant "Forward Look" offered by Chrysler Corporation for 1955-56 hurtled into '57 with enough aesthetic force to turn Detroit auto design on its ear. What had been the stodgiest of the Big Three was suddenly the industry's design leader. Company offerings for 1957 were delta-shaped in profile: low, lean, and swoopy, and although they gave the impression of great length, they actually were a bit shorter than the '56 models.

In a Detroit era when show cars had names like Predictor and Nucleon, the various new Chrysler products really did have an aura of futurism. A Plymouth ad line shouted, "Suddenly, it's 1960!"—and it sounded like understatement.

For the period, the Chrysler vehicles were restrained and carried the new look well. Form didn't exactly follow function, but the great fantails of fins didn't appear tacked on, either, and chrome and side trim were reasonably simple as well.

At Chrysler Division, the New Yorker had been the top of the line off and on since just before World War II, and rode that status into the late Fifties because the Imperial that had top billing at times was spun off as its own marque for 1955.

For 1957 and '58, the New Yorker ran with a 392-cid "hemi" V-8 producing 325 or 345 bhp. The hemi had a heavy mystique, but it was expensive to produce, so for 1959 Chrysler fitted the New Yorker with a 413-cid V-8 sporting wedge-shaped combustion chambers, a four-barrel carburetor, 4.18×3.75-inch bore and stroke, and a 10.0:1

compression ratio. The motor lacked only the hemi allure; in power, it was that engine's equal, and more, producing a thumping 350 horses at 4600 rpm and 470 pound-feet of torque at 2800. The sole transmission was a three-speed TorqueFlite automatic.

Front suspension brought upper A-arms, lower traverse arms, longitudinal torsion bars, and an antiroll bar. Out back was a live axle and semi-elliptic leaf springs. Brakes, front and rear, were drums.

The '59 hardtop coupe had some physical substance, weighing in at 4080 pounds but able, nonetheless, to sprint from zero to 60 in about 10 seconds. Top speed was 115 mph.

On the highway, these cars were smooth, powerful performers with spacious, airy interiors, and gave away little to Cadillac, Lincoln, or Imperial. But most every design element became a bit more baroque for 1958 and again for 1959—by which time Ford and General Motors had introduced imposing, swoopy cars of their own.

In 1959, a New Yorker hardtop coupe went out the door for $4476—up from $4347 in 1958 and $4202 in '57. Unfortunately, indifferent workmanship and a tendency to rust meant that the so-so sales figures for the 1957 Chrysler line only got worse for '58 and '59. Some 8863 New Yorker hardtop coupes were produced for 1957. For '58, the total fell to 3205 (to be fair, this was due partly to the year's serious economic recession) and suffered another tumble for '59, when just 2435 units were produced.

# 1959 JAGUAR
## MARK IX FOUR-DOOR SEDAN

During World War II, Jaguar decided to develop a dohc engine that would power sedans capable of 100 mph. Only race cars and exotics such as Duesenberg, Bugatti, and Alfa Romeo had used such a valvetrain before World War II. The twin-cam's efficient breathing allowed Jaguar's XK engine to wring 160 bhp from a 210-cid six—the same power as Cadillac's 331-cid V-8 introduced at about the same time.

Because the XK was designed to serve sedans, it was also smooth, silent, and durable. Jaguar made sure the engine looked impressive with polished-aluminum cam covers. It would be in production for 44 years.

In order to ramp up XK engine production before volume use in the sedans that paid the company's bills, the first car to get the powerplant was the XK120 sports car introduced at the 1948 London Motor Show. For 1951, Jaguar introduced the XK-powered Mark VII sedan. Styling was of the stately English school, but the Mark VII lost the prewar look of earlier Jaguar sedans. Inside, though, there was still as much wood and leather as you would find in a prewar Rolls-Royce or Bentley.

The Mark VII shared the XK120's torsion-bar independent front suspension and had good handling for a 3700-pound sedan. It had a top speed

CALIFORNIA
SCRLET 9
MK IX

Jaguar
Automatic

of 101 mph and did well in rallies.

Jaguar hoped that a large (by British standards), comfortable sedan with good performance would do well in the important American market. One American impressed by the Mark VII was *Road & Track* editor Bill Corey. Corey took a Mark VII on a road trip in '52 and wrote, "[W]e cruised the car at slightly under 'red-line' on the tach, about 100 miles per hour. There is no question that this is the greatest highway car your author has ever had the pleasure of driving." Corey went on to say, "[S]ome of the detours we encountered defied description. They offered a chance to break the heart of the Jag, and try we did. We drove 70 [mph] over roads that other cars could scarcely negotiate at 25. Nothing broke. Nothing happened. Not a squeak nor a rattle. Nothing."

A Mark VIII was introduced in 1956 with more power and revised styling. The Mark IX of 1959 looked virtually the same as the Mark VIII, but had several important improvements. The XK engine had grown to 231 cid (3.8 liters) and horsepower was up to 220. Top speed was now 114 mph with 0-60 mph in 11.3 seconds. To handle the extra power, Mark IXs gained four-wheel disc brakes. Power steering was standard and must have helped sales in the U.S., as did the Borg-Warner three-speed automatic transmission—standard on cars bound for America. The Mark IX was in production through 1961, when it was replaced by the longer, lower Mark X.

This car retains its factory tool kit. Perhaps as a comment on Jaguar reliability, an unusual collection of tools and spares was mounted in recessed compartments in the doors, including light bulbs, a spark plug, and a grease gun.

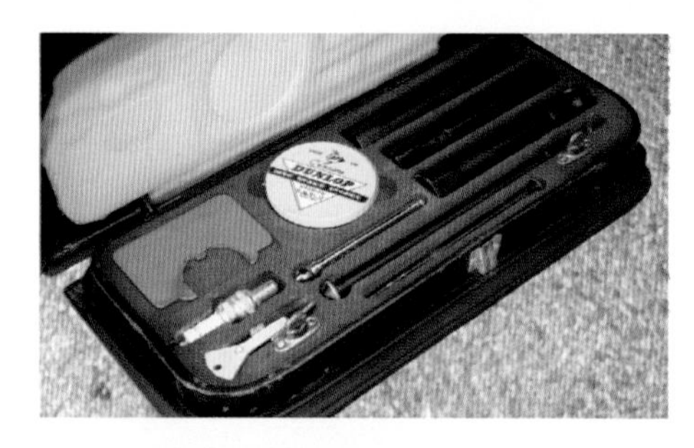

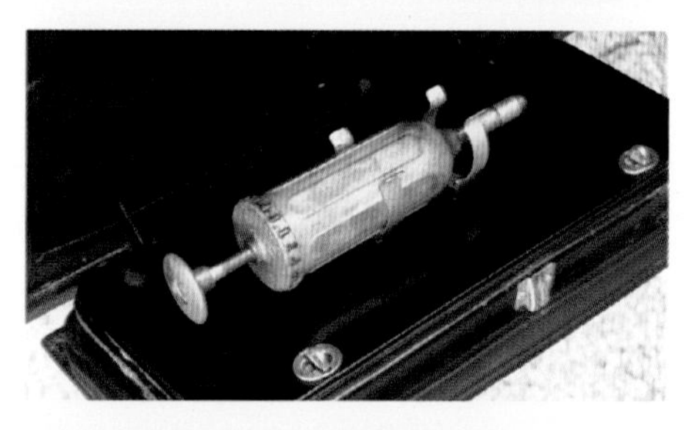

SCRLET 9

CALIFORNIA
SCRLET 9

# 1960 CADILLAC
## ELDORADO SEVILLE HARDTOP COUPE

Cadillac's top-line Eldorado convertible was selling so well in the mid Fifties that Cadillac added a two-door-hardtop companion for '56. The convertible gained the double-barreled title Eldorado Biarritz to distinguish it from the new Eldorado Seville hardtop.

Cadillac had found a new niche and the coupe handily outsold its convertible sibling the first year out. Hardtop and soft-top Eldorados were the same price each year that they were available, and the hardtop continued to outsell the convertible until 1959.

Eldorados had unique sheetmetal in some years, but the 1959 and '60 models had the same body panels as other Cadillac coupes and convertibles. It was up to distinctive side trim and a unique "grille" beneath the decklid to differentiate Eldorados from the base models. The Seville also had a top covered in vinyl-coated fabric. Inside, Eldorados had richer interiors.

Cadillac's trademark tailfins reached their peak in 1959, but tastes were changing and the fins were trimmed back for 1960. Leading the charge for

ELDORADO

Cadillac
GENERAL MOTORS CORP. DETROIT, MICH
Air Conditioning

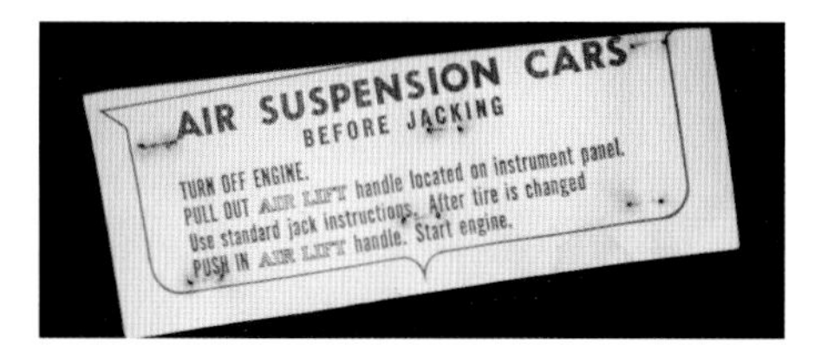
AIR SUSPENSION CARS
BEFORE JACKING
TURN OFF ENGINE.
PULL OUT AIR LIFT handle located on instrument panel.
Use standard jack instructions. After tire is changed
PUSH IN AIR LIFT handle. Start engine.

cleaner, crisper styling was General Motors styling chief Bill Mitchell. Mitchell took over GM styling with the retirement of the famed Harley Earl in 1958. Under Mitchell's direction, Cadillac fins would shrink and its styling become more graceful and restrained.

The standard Cadillac engine was a 390-cid V-8 with a four-barrel carburetor that put out 325 bhp. Eldorados replaced the four-barrel with three two-barrel carbs that raised horsepower to 345. Cadillac introduced air suspension on its limited-production Eldorado Brougham sedan in 1957. It was standard on the Biarritz and Seville for 1958-60 and optional on other Cadillacs. Air suspension was considered a technological wonder when introduced, but proved troublesome and was dropped after 1960. Many 1957-60 Cadillacs with GM's air suspension have had their air bags replaced by conventional steel springs, but this car still rides on air.

The Eldorado Seville was also dropped after '60. The convertible Eldo outsold the coupe that year and all Eldos would be convertibles until a distinctive front-wheel-drive Eldorado coupe was brought out for 1967. The Seville name would reappear with the introduction Cadillac's compact, import fighter in 1975.

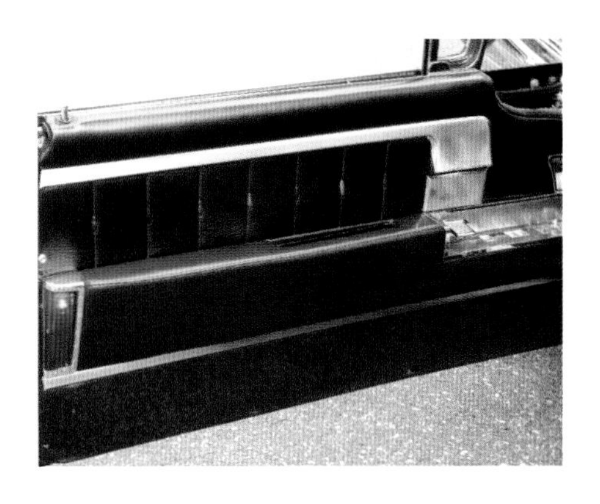

CRVISE CONTROL
OFF
TO CRUISE

# 1961 CHEVROLET
## CORVETTE CONVERTIBLE COUPE

Eight years removed from the Corvette's debut, the '61 model was still part of what's now called the "C1" generation—essentially 'Vettes with a straight rear axle. An all-new, flowing "ducktail" rear end, borrowed from Bill Mitchell's Stingray racer, was the 1961 Corvette's biggest change. It added luggage space as well as visual appeal. Small round taillights sat alongside the central license plate recess, while a modest creaseline extended through the trunklid's traditional round medallion. Simple bumperettes were mounted below the taillights.

Up front, body-colored bezels helped clean up the quad-headlight nose. In addition, Corvette's trademark grille teeth were removed and replaced with a horizontal-mesh insert.

There was little different under the hood, save for a running change to a cross-flow radiator. The base powerteam that came in the $3934 starting price was a 230-bhp 283-cid V-8 hooked to a three-speed manual transmission. A four-speed manual and Powerglide automatic were alternate transmission choices. Twin-carburetor engines of 245 or 270 bhp were available, as were fuel-injected variants of 275 or 315 horsepower. Powerglide was not offered with the three hottest engine choices, and most buyers opted for the extra-cost 4-speed over the base manual. A 315-horse "fuelie" with the four-speed manual transmission could go from 0-60 mph in 5.5 seconds and top 130 mph.

Seven exterior colors were available, and the featured car is sprayed in one of the year's new ones: Fawn Beige. It is complemented with Ermine White side

coves. A Corvette fixture since 1956, the contrast-color coves would not return to the options list after '61. The extra-cost wide-whitewall tires were also putting in their final appearance this year.

The matching Fawn interior was one of three possible colors that could be paired with this paint. Passengers who settled into the two vinyl bucket seats had a little more room to themselves thanks to a newly narrowed transmission tunnel. Windshield washers and sun visors were among a handful of new standard items.

Advertisements for the 1961 Corvette tended to concentrate on "lifestyle" and driving enjoyment, but performance remained a vitally important attribute. Racing success continued at Sebring, Florida, with Corvettes taking the top three places in their class at the annual

12-hour race. It was an impressive showing considering these were near-stock Corvettes, and the top-placed car finished 11th overall against much more expensive and exotic machinery. Another Corvette won its class at the Pike's Peak hillclimb in Colorado.

Corvette's slow, steady sales climb continued in 1961. The 10,939

made—all convertibles—represented a 6.6-percent gain from 1960.

The featured car runs the standard engine with a single Carter four-barrel carb and the three-speed transmission. Options include an AM signal-seeking radio ($137.75), removable hardtop ($236.75), two-tone paint ($16.15), and whitewall tires ($31.55).

CORVETTE

# 1962 MERCURY
## MONTEREY S-55 HARDTOP COUPE

Full-sized cars with sporty bucket seats and center consoles were all the rage in the early Sixties, with Pontiac's Grand Prix and Chevrolet's Impala Super Sport grabbing market niches. Mercury entered the field in mid 1962 with the Monterey S-55. Just as mainstream Montereys were related to the Ford Galaxie, the S-55 was Merc's version of the bucket-seat Galaxie 500/XL.

The Monterey S-55 was available as a $3488 hardtop coupe or $3738 convertible. Both rode on a 120-inch wheelbase and had a standard 292-cid V-8 of 170 bhp. Extra-cost upgrades included a 352 V-8 good for 220 bhp, 390s that generated 300 or 330 horses, and two new Marauder 406-cid V-8s with 385 or 405 ponies—the latter with triple two-barrel carbs. A Multi-Drive three-speed automatic transmission was standard; a Borg-Warner four-speed manual was optional. (Cars with either of the 406 V-8s required the four-speed and came on a 119-inch chassis with heavy-duty suspension parts.)

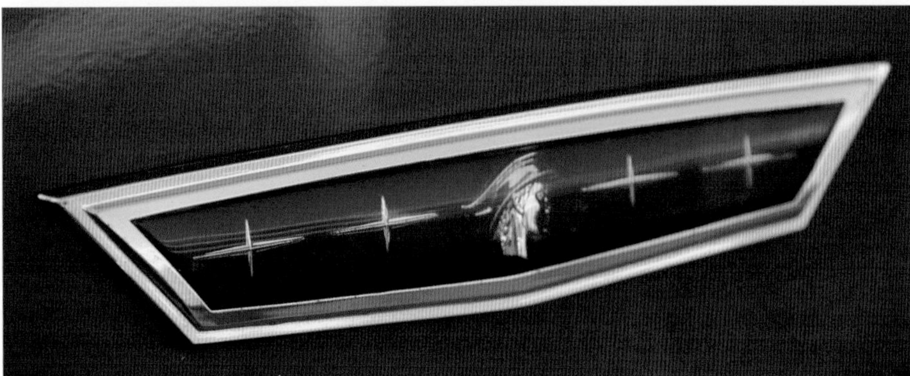

The S-55 featured unique exterior trim and a special vinyl-upholstered interior. A floor shift sprouted from the brightly trimmed console regardless of transmission choice. Only 4087 S-55s were built during its short introductory year: 2772 hardtops and 1315 convertibles.

The 1962 Mercury Monterey S-55 hardtop that's featured here may look restored, but it's an original car with 86,000 miles on the odometer. The first owner drove this car until the early Eighties when poor health forced him to store it in a garage. There it sat until 2000. The car was in excellent condition and only a good polishing was required to bring the paint back to like-new condition. The special S-55 wheel covers with distinctive red, white, and blue centers are original. The first owner always kept them in the trunk. As usual for a car awakened from a long sleep, the S-55 needed new tires, a carb rebuild, and brake-seal replacements; otherwise, it was mechanically sound.

FUEL
TEMP

1963
BUICK
RIVIERA
SILVER ARROW I

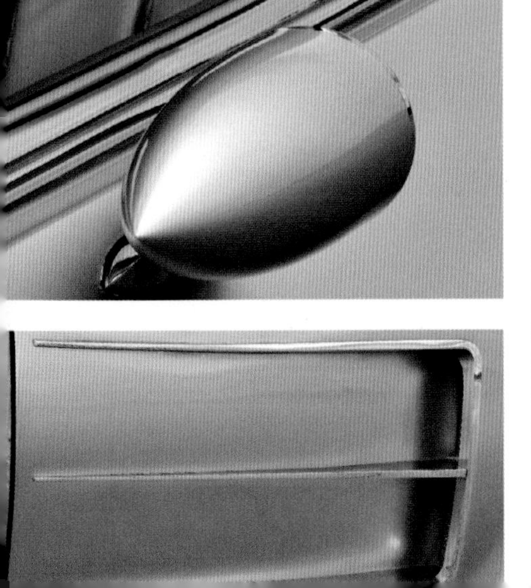

With rare exception, the personal-luxury Rivieras made from 1963 to 1999 were special cars for Buick. When introduced, the "Riv" gave a stylish shot in the arm to a marque that had lost its luster in the late Fifties and was fighting to win it back. The very name became as synonymous with Buick as portholes.

However, some Rivieras were designed to be more special than others. A case in point is the 1963 Riviera Silver Arrow featured here. It was a show car, yes, but it was also the personal transportation of one of the auto industry's master showmen, General Motors Vice President of Styling William Mitchell.

The Riviera was among the first defining successes in Mitchell's tenure as vice president since succeeding his mentor, Harley Earl, in late 1958. He wanted a Ferrari crossed with a Rolls-Royce that could challenge the Ford Thunderbird in the burgeoning personal luxury market, and that's what he got.

As proud of the production car as he was, Mitchell sought something distinctive for the one he was going to drive. He started by chopping the top by more than two inches and slightly recontouring the roof. Door glass was ventless, predicting the new look that would arise from the adoption of flow-through ventilation systems.

In another nod to the Riviera's future, Mitchell put the headlights behind translucent caps in the fender ends. Production '65 Rivs would have their lights behind retracting doors in this same location. The bodyside vent was modified to have more of a pitchfork look, and ribbed rocker-panel trim was another preview of '65. Sleek body-color racing-type mirrors were fitted, too. In back were distinctively trimmed taillights, with brake lights fitted into slits in the "tulip panel" below the backlight.

The interior was redone with special deeply contoured bucket seats front and rear. Cabin decor was topped off with distinctive silver leather upholstery and matching floor mats.

The Silver Arrow initially had a stock-looking eggcrate grille and wire wheels. However, running modifications brought a pointed nose more than three inches longer and grille shutters that open or close automatically in response to the temperature of the 425-cid Buick V-8. Concentric "classic" wheel covers and wide whitewalls were added, too.

In 1968, Buick had to start numbering Silver Arrows like Super Bowls. A Silver Arrow II was whipped up from the then-current design. The II reportedly was an internal styling study and never put on public display. However, in 1972, auto-show visitors were exposed to the Silver Arrow III, a feature-packed customization of the polarizing "boattail" Riviera in Buick showrooms at the time. The name's final use came with the final Riv: 200 fully optioned 1999 models were painted a unique shade of silver and badged as Silver Arrows.

# 1963 FERRARI
## 400 SUPERAMERICA COUPE SPECIALE AERODINAMICA

At the urging of Luigi Chinetti, his distributor in the U.S., Enzo Ferrari agreed to build a powerful and exclusive tourer with the American market in mind. The result was the 410 Superamerica, made in three series between 1956 and 1959.

A new Superamerica—the 400—made its formal debut in January 1960 at the Brussels Auto Salon. Its Pininfarina-designed cabriolet body had lines that called to mind the Ferrari 250 GT. Its 95.3-inch wheelbase was shorter than that of the 410, but the new car's chassis included telescoping shock absorbers (in place of the former lever-action shocks) and Dunlop disc brakes at all four wheels. Under the hood sat a new 4.0-liter "short-block" V-12 engine that traced its lineage to a design penned by engineer Giocchino Colombo in the late Forties.

The development of the 400 Superamerica took another leap forward when Pininfarina showed a sleek coupe, dubbed "Superfast II," at the Turin show in November '60. With modifications, it went into production as the Coupe Speciale Aerodinamica, 13 of which were constructed in 1961-62.

Visitors to the London Motor Show in October 1962 were the first to see a revised 400 Superamerica. Its major distinction was a wheelbase stretch to 102.4 inches. The added length opened up additional luggage space inside the two-seat coupe.

The Aerodinamica coupe seen here is one of the 19 built on the longer chassis before series production ended in 1963. (Total 400 Superamerica output came to 48 cars.) Its ohc V-12 is fed through three two-barrel carburetors and reaches its peak 340 bhp at 7000 rpm. The transmission is a fully synchronized four-speed with an electrically activated overdrive. The single dry-plate clutch is another of the details that separates the 400 Superamerica from the 410. The top speed was around 160 mph.

Originally built for a member of the Rockefeller family, the car features a low, elliptical grille and covered headlamps. (The cars were built to customer specifications and some have slightly different bodywork and equipment.) A dashboard knob activates radiator shutters that could be used to aid engine warmup on cold days.

Soon after its arrival in the U.S., the car's color was changed to silver. However, it has been returned to its original hue.

# 1963 STUDEBAKER
## GRAN TURISMO HAWK

The Studebaker name persisted in new-car showrooms until 1966, but the end of automaking as the company had known it for decades really came in 1963. A steady decline from the production peak in 1950 had gained an unmitigated momentum by the time the Sixties dawned. When Studebaker lost $25 million on automotive operations in '63, the corporation closed its main factory in South Bend, Indiana, and shifted assembly of only its highest-volume car line to a smaller, cheaper-to-run plant in Canada to live out the brand's last few years.

When the end came in South Bend near Christmas 1963, Studebaker was, of course, busy producing its 1964 models. It was a varied lineup of products, headed by the dramatic fiberglass-bodied Avanti coupe, and it also included the almost-intermediate

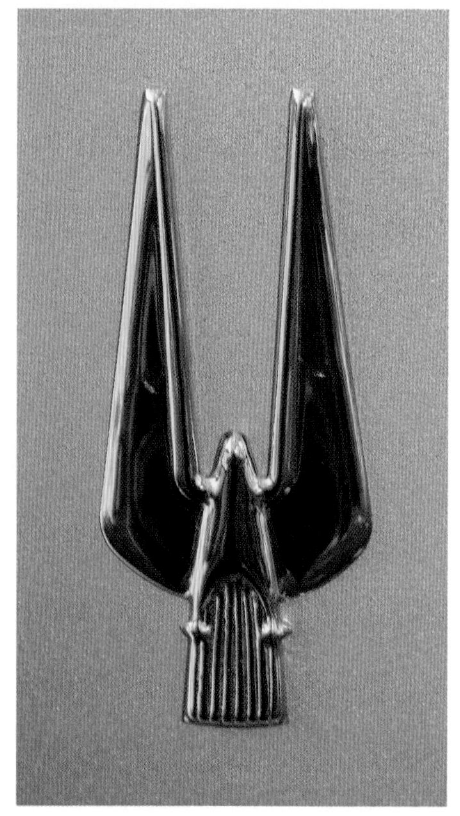

compact Lark (a name Studebaker was starting to downplay), the personal-luxury Gran Turismo Hawk hardtop coupe, and trucks. Only the family friendly Larks could be spared in Canada. The Avanti and GT Hawk were abruptly halted with '64 production of just 809 and 1767, respectively.

The Avanti would rise again in a few years as an independent marque, but for the Hawk, this was the end of an odyssey that began 11 years earlier. At its heart, the GT Hawk was the brilliant Starlight coupe/Starliner hardtop designed for 1953 under the auspices of Raymond Loewy. In '56, when restyled sedans and wagons were ushered in, the sleek 120.5-inch-wheelbase two-doors were dolled up with trendy fins and chrome and renamed Hawks.

Hawk coupes were still in the line in 1961, but their looks were by then

Gran Turismo

TEXAS
63 HAWK

decidedly old-hat. New president Sherwood Egbert called on industrial designer Brooks Stevens to come up with something new for '62, albeit on the same platform. With just $7 million at his disposal (some of which had to be used to touch up the Lark) and only a few months to work, he lopped off the fins, toned down the trim, and applied a hardtop roof—the Hawk's first since '58—clearly inspired by the successful look of the period Ford Thunderbird and Galaxie.

Sales more than doubled to 8388, but started to dip again in '63, when just 4634 of the $3095 cars were ordered. A revised grille surface, round parking lights, and body-color headlight rings were subtle appearance updates. More substantive were the addition of

optional 240-bhp R1 and 290-bhp supercharged R2 versions of the Studebaker 289-cid V-8. *Motor Trend* reported the R2 would push a 3780-pound GT Hawk with the performance-oriented "Super" package to 60 mph in 8.5 seconds and run the quarter-mile in 16.8 seconds.

This car runs the standard Hawk engine, a 289 with a two-barrel carburetor, rated at 210 bhp. It is hooked to a three-speed manual transmission with overdrive. Extras include power steering, power brakes (discs in front), a radio, and air conditioning.

## 1966 FORD
THUNDERBIRD CONVERTIBLE

The first Ford Thunderbird was a two-passenger sports car, but for 1958, T-Bird became a four-passenger "personal luxury car".

With a nod to the T-Bird's roots as a two-seater, this 1966 Thunderbird was equipped with a very rare accessory fiberglass tonneau. Sloping headrest fairings trailed back from the front seats over the removable tonneau, which,

when in place, covered the curved, loungelike back seat. Ford made 5049 Thunderbird convertibles for 1966, but precious few with the tonneau. The tonneau first appeared on the Thunderbird Sports Roadster, which had been a full-fledged model in 1962 and '63. The Sports Roadster cost $650 more than the standard convertible and never caught on. Waning sales doomed the Sports Roadster before the fourth-generation 'Bird made its debut in 1964, but a restyled tonneau popped up as a $269 item on the accessory list. Just a handful are thought to have been ordered for the roughly 21,000 Thunderbird ragtops made from 1964 to '66. Sports Roadsters also came with genuine wire wheels; a set was ordered for this car to complete the look.

Nineteen sixty-six was the final year for the fourth-generation Thunderbird, so changes were few. They were, however, noticeable.

The frontal facelift, carried out under the direction of L. David Ash, did away with the bumper/grille look of 1964-65. In its place were a larger eggcrate grille above a simpler blade-style bumper and body-color valance panel.

A big Thunderbird emblem spread its wings across the center of the grille. The blunt, slightly raised faux hood scoop of prior years became a subtler bulge with a vee'd front.

Bodysides were cleaned up with the deletion of the simulated scoops seen on the front fenders of the '65s. Rear-fender skirts became options. A full-width taillight ensemble (with a central back-up lamp) replaced a twin-lamp design, but turn signals still blinked sequentially from the middle out, an attention-grabbing gimmick first seen in 1965.

A 428-cid V-8 good for 345 bhp joined the options list, providing T-Bird buyers with their first choice of engines since '63. The standard powertrain remained Ford's trusty 390-cube V-8 and the Cruise-O-Matic three-speed automatic transmission, but thanks to a half-point compression boost, the base mill gained 15 horsepower to 315 bhp. This restored car is powered by a 390 dressed up with a chromed air cleaner and valve covers. *Car Life* tested a 1966 Thunderbird with a 390 V-8 and achieved a top speed of 115 mph.

For 1966, base prices declined a bit from 1965—by $74 in the case of the convertible, to $4879. But sales retreated, too, by almost eight percent. All-new Thunderbirds would arrive for 1967, but a convertible wouldn't be among them.

Thunderbird

Thunderbird
ROKNROL6

# 1967 BMW

## 2000C HARDTOP COUPE

BMW as we know it today began with the "New Class" 1500 introduced in 1962. The 1500 was a compact four-door sedan with a 100.5-inch wheelbase. The chassis was a clean-sheet design. Up front, there were MacPherson struts. Out back, an improved version of the semi-trailing arm and coil-spring independent rear suspension that had originally been designed for the BMW 600 and 700 was used. Other mechanical items included worm-and-roller steering and power-assisted front-disc/rear-drum brakes.

The 1500 was powered by a 1.5-liter ohc four-cylinder engine—the M10—that liked to rev and also proved durable. It was rated at 75 horsepower and mated to a 4-speed manual transmission.

The 1500 was well-received in Germany, and the county's automotive press thought it filled the market niche once

occupied by the well-regarded Borgward Isabella.

BMW's U.S. distributor, Max Hoffman, started imports in time for the 1963 model year. The East Coast port-of-entry price was $3550. *Road & Track* tested an early example for the September 1963 issue and concluded, "No car is perfect, but in the case of the BMW 1500, the effort of its designers and builders to approach perfection as nearly as possible within their limitations is pleasingly evident."

BMW expanded on its success with more powerful 1800 and 2000 variations on this winning theme. Then a shorter, lighter 2002 two-door sedan came out with even better performance than the four-door, and established BMW as the builder of "ultimate driving machines." The 2002 gave way to the iconic first-generation "E21" 3 Series in the mid Seventies.

In between these two developments, a graceful 2000 coupe was introduced during 1965. The 2000 shared the sedan's mechanical components and wheelbase, but had its own sheetmetal, which was formed by coachbuilder Karmann. The 2000 was designed for style and luxury rather than outright performance, and was actually 65 pounds heavier than the sedan. It used the largest 1990cc version of the M10 engine, and was offered in two states of tune. The 2000C used a single Solex carburetor and developed 100 bhp, while the 2000CS used dual Solex carbs for 120 bhp. The C had a top speed of 105

mph; the CS was good for 115. The sportier CS also had a standard roll bar that was optional on the C.

BMW added a more powerful six-cylinder version of the coupe in 1968. It immediately outsold the 2000C/CS fours, which were dropped after 1969. Today, the 2000 coupe is a rare car in the U.S.

The restored 1967 2000C on these pages wears flush European-style headlights rather than the exposed sealed-beam quad headlights originally required on American-market cars.

Lincoln rebooted its styling themes several times in the Fifties, but couldn't break Cadillac's postwar dominance of the luxury market. Ford Motor Company executives concluded that this lack of design consistency was hurting Lincoln. It was decided that stretching the design cycle to as long as nine years would establish a Lincoln "look," reduce body tooling costs, and increase resale values.

In the late Fifties, Ford Vice President Robert McNamara wanted to kill Lincoln. McNamara was extremely practical and tight-fisted with the company funds. The flamboyant Lincoln division had been losing money, and McNamara had no sympathy for the fabled brand. However, McNamara was persuaded to give Lincoln one more

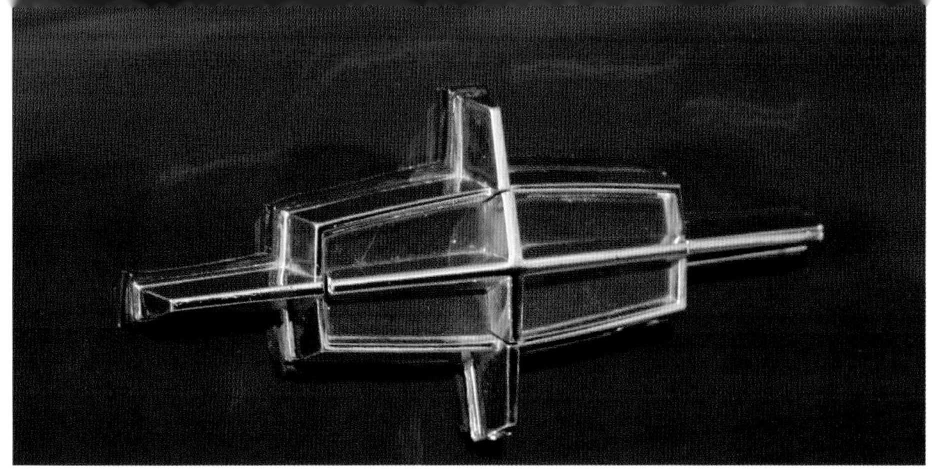

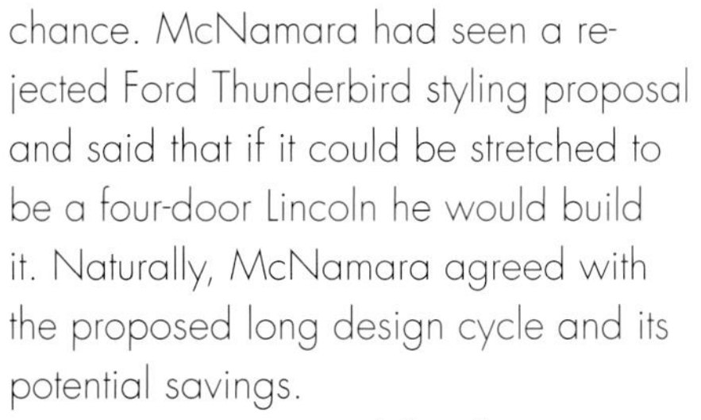

chance. McNamara had seen a rejected Ford Thunderbird styling proposal and said that if it could be stretched to be a four-door Lincoln he would build it. Naturally, McNamara agreed with the proposed long design cycle and its potential savings.

McNamara insisted that the next Lincoln design be reasonably sized. A short (by luxury-car standards) 123-inch wheelbase was chosen. A problem with the short body was that it was hard to exit the back seat without kicking the door. That was solved by having the rear doors hinged at the rear. "Suicide doors" were a Lincoln Continental feature up to 1970, even after the wheelbase was lengthened to 126 inches in '64.

The classically simple 1961 Lincoln lacked design clichés that would go out of style. Thus, the Continental sold and aged well. In fact, Lincoln earned a $20 million profit between 1961 and '64 and saved the brand from extinction.

By '66 Lincoln was ready for a facelift. The 1966 Lincoln was an updated version of the successful 1961 design. It was subtly different, but carried on the crisp appearance of the original. With yearly detail tweaks, the 1966 form continued through 1969.

Lincoln worked to improve quality with the 1961 Continental, and claimed that it took four days to build a car. Two of those days were spent assembling the unitized body. The cars were meticulously assembled and subjected to a 90-minute inspection before being approved for delivery.

This remarkably well-preserved 1968 Lincoln Continental is a testament to Lincoln quality. It was repainted, but the cloth upholstery is original, as is the spare tire. Very little mechanical work was required during this car's long life.

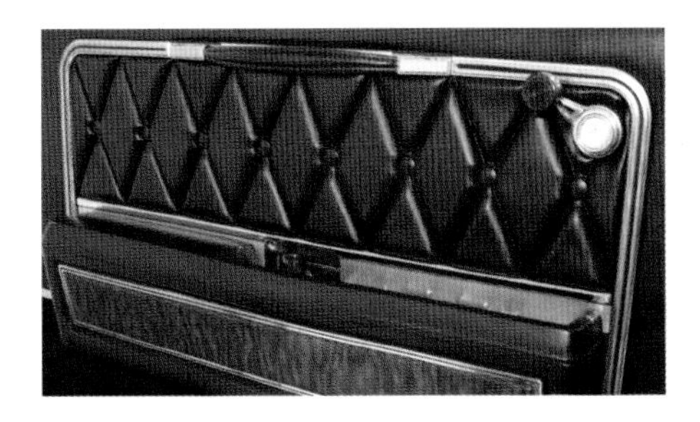

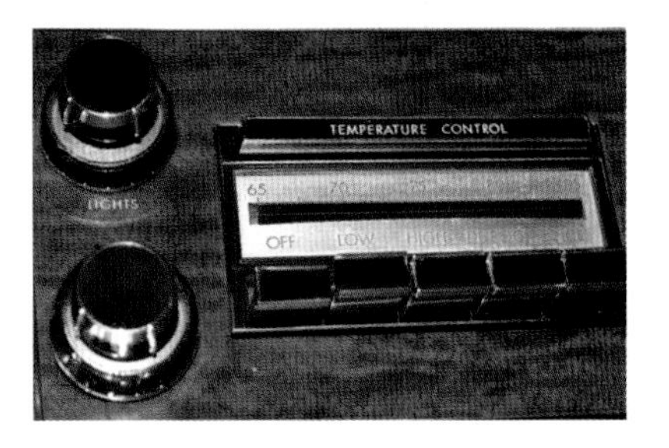

# 1970 JAGUAR
## XJ6 FOUR-DOOR SEDAN

Jaguar's XJ is one of the longest-running model names in the industry. The XJ made its debut in September 1968 as a 1969 model, and Jaguar's flagship sedan is still badged XJ today.

There is good reason why a Jaguar sedan launched in the late Sixties should give birth to such a long-lived name. At its introduction, the XJ set a new standard for European luxury sedans. The press raved. In the UK, *Motor* said, "We believe that in its behavior it gets closer to overall perfection than any other luxury car we have tested, regardless of price. The car just floats round

corners with such enormous reserves of adhesion that the driver's nerve will invariably be lost before the grip." In the U.S., *Road Test* exclaimed, "The Jaguar XJ6 also just happens to be the best riding car we have ever driven, and its ride/handling ratio is not equaled by any other car we can think of." *Road & Track* deemed the XJ "uncannily silent, gloriously swift, and safe as houses."

Many of the components of the new XJ were familiar. The XK 4.2-liter six-cylinder engine had been launched in 1948. It was an advanced dohc design that had powered five Le Mans winners and yet was smooth and durable. Emissions standards were starting to cut the power of the 20-year-old design, but its 245 bhp provided good performance—*Road & Track* claimed the top speed

was 124 mph, with 0-60 mph in 10.1 seconds. Its independent rear suspension had been introduced in the E-Type and Mark X in 1961.

Meanwhile, the body was all new. It used unitary construction with the suspension carried by rubber-mounted sub frames to isolate road shock and vibration. The firewall was double-skinned to keep engine noise out of the cabin.

Just as the XJ's engineering received great acclaim, styling also drew praise. Jaguar's founder, Sir William Lyons, gained renown for his styling of the XK120 and E-Type sports cars. The XJ was perhaps his best sedan design. The look of the original XJ was so successful that Jaguar used variations of the theme through 2009. Inside were the wood and leather expected of a Jaguar. The dashboard followed the E-Type layout with a comprehensive array of gauges and a long row of rocker switches.

Demand was high for the XJ6, but production was slow to get moving, and labor unrest was a constant problem. In England, the wait for delivery was soon more than 12 months. Indeed, supply and reliability would be persistent challenges for the XJ.

Still, the inherent good qualities of the XJ kept the original design in production through 1992. Along the way, a V-12-powered XJ12 was added. There were Series II and Series III updates as well. A totally new XJ6—albeit with styling that echoed the original—came out in 1986. However, its engine bay was too narrow to take the V-12, so the Series III XJ12 remained in production until 1992.

Our featured 1970 Jaguar XJ6 is from the first year of American imports. It is equipped with a Borg-Warner three-speed automatic transmission, which came standard on U.S.-bound XJs.

TEXAS 72
NCR·281

# 1972 OLDSMOBILE
## NINETY-EIGHT REGENCY HARDTOP SEDAN

Both Oldsmobile and Cadillac celebrated anniversaries in 1972, Cadillac its 70th and Oldsmobile—then the oldest surviving American automaker—its 75th. Both decided it was the velour anniversary. The fabric had vanished from car interiors after World War II, but thanks to the lead of Olds and Cadillac, velour became the number one upholstery choice for American cars in the Seventies.

Oldsmobile's velour seat had a pillowed effect that made it look like an extremely inviting couch. This plush-sofa look was also widely copied during the Seventies. Olds had just created the exuberant luxury that Americans of that era craved.

This trendsetting interior made its debut in the Ninety-Eight Regency four-door hardtop, a midyear model with a production run limited to 5000 copies for '72. Tiffany & Co., the New York jeweler, styled the face of the Regency's dashboard clock and provided a sterling-silver key ring. Every Regency was sprayed in Tiffany Gold metallic paint, but the interior was available in a choice of black or gold.

Olds realized it was on to a good thing and brought back the Regency as a regular model for '73, selling more than 34,000 of them. The name would remain a part of the Ninety-Eight family until the end of that line in 1996, after which it transferred to the Eighty-Eight range through '98.

John Beltz, Oldsmobile's general manager from 1969 to 1972, said Olds buyers didn't want small cars, and the Ninety-Eight certainly wasn't, not with

a 127-inch wheelbase and an overall length of 228 inches. Beltz also said it was much easier to reduce emissions in a big engine because it was usually operating at only part throttle. Of course, he said this before the 1973-74 OPEC oil embargo, when most Americans weren't yet overly concerned with fuel economy. The Ninety-Eight had a 455-cid 250-bhp V-8 to move its 4698 pounds, and it moved them well. A '71 Ninety-Eight did 0-60 mph in 8.7 seconds (but averaged only 11 mpg) in *Motor Trend* testing.

The Regency pictured here was purchased new by an elderly lady who drove it only 12,288 miles and kept it in an air-conditioned garage. Thanks to gentle use, it remains in like-new condition. Only tires, battery, and belts have been replaced. The rest is original. Options include an eight-track tape player and an external temperature gauge mounted outside on the driver's door.

TEXAS
NCR-281

120
110
100
80

Ninety Eight

# 1992-2000 LEXUS
## SC 300/400

As the story goes, Toyota designers were told to make the first Lexus SC such a beautiful sport coupe that it would be a recognized collector car 30 years hence.

Whether they are or not collectible, these cars had much to offer. Like most every Lexus, the SC 300 and SC 400 were beautifully crafted, long-distance comfortable, luxurious, high-tech sophisticated, and very reliable. It was that formidable blend of assets that helped Lexus unseat BMW and Mercedes-Benz as America's top-selling luxury nameplate in less than 10 years

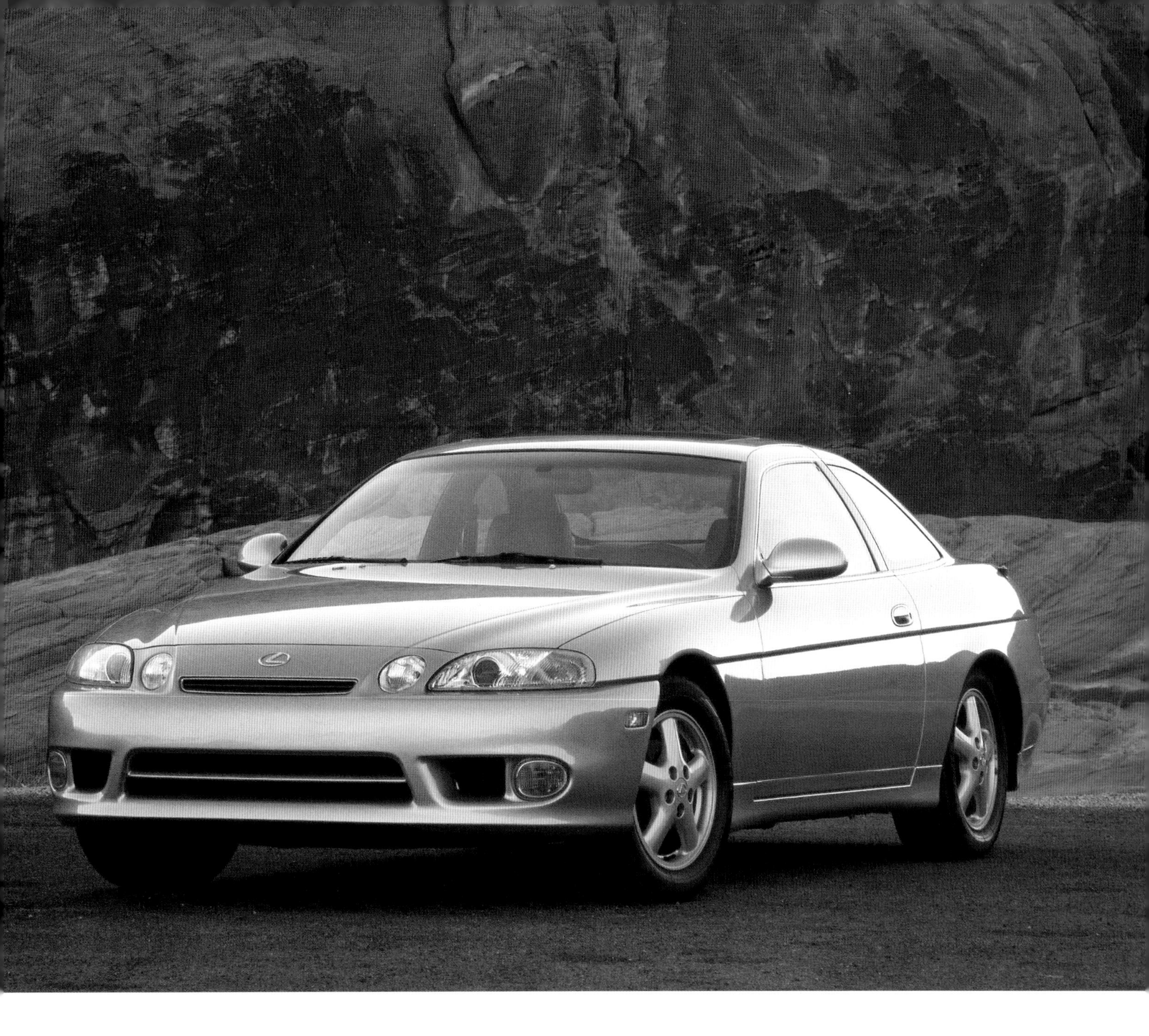

while winning a slew of quality awards from the likes of J. D. Power. Moreover, despite the Lexus reputation for floppy handling and numb steering, these SCs were rewarding to drive. They may not have been the quickest or most nimble things on four wheels, but they did most everything well and provided much pleasure in the bargain.

Six-cylinder and V-8 engines were planned from the start. The former was a 3.0-liter inline unit similar to the contemporary Toyota Supra six, while the latter was the new 4.0-liter LS sedan motor. Both were all-aluminum designs with twin overhead camshafts operating four valves per cylinder. The six delivered 225 bhp and 210 pound-feet of torque. The V-8 claimed 250 bhp and 260 pound-feet. A four-speed automatic transmission was standard in the 400 and available for the 300 in lieu of a five-speed manual. In 1998, SC borrowed powertrains from Lexus' redesigned GS midrange sedans. That brought SC 400 outputs to 300 bhp and 310 pound-

feet of torque. A five-speed automatic was also new for the V-8 model. The SC 300 stayed with a four-speed automatic, but lost its manual gearbox to lack of interest, though another 10 pound-feet of torque—220 in all—was some consolation.

Most chassis hardware was also LS, but tailored for a sportier drive. Despite all the parts-sharing, the SCs had their own unibody structure with tidier dimensions than the LS sedan. Wheelbase was 4.9 inches shorter at 105.9. Lexus said the SC 400 weighed 3575 pounds at the curb, 184 less than the LS. The SC 300 scaled 3494 pounds.

The SCs arrived almost dead-center of the luxocoupe field, and not just in size. The 300 opened at $32,000 with automatic, the 400 at $37,500, both very competitive figures. The SC faced competition from Cadillac Eldorado, Lincoln Mark VII, Jaguar XJS, Mercedes E300, and BMW's very pricey 850i.

Performance stats were surprisingly consistent through the years and surprisingly good for luxury cruisers, especially the SC 300. *Car and Driver* reported 0-60 mph at 6.8 seconds with the six (and manual shift), 6.7 for the V-8, plus equally close skidpad pull of 0.85g versus 0.86. *Road & Track's* testing yielded somewhat wider gaps, with respective 0-60 times of 7.4 and 6.9 seconds, plus lateral acceleration of—wait for it—0.89g versus 0.86. Lexus claimed 6.6 seconds 0-60 for post-1997 SC 400s. Plus, the SC 300 and SC 400 were about as pleasant and worry-free as potential collectibles get.

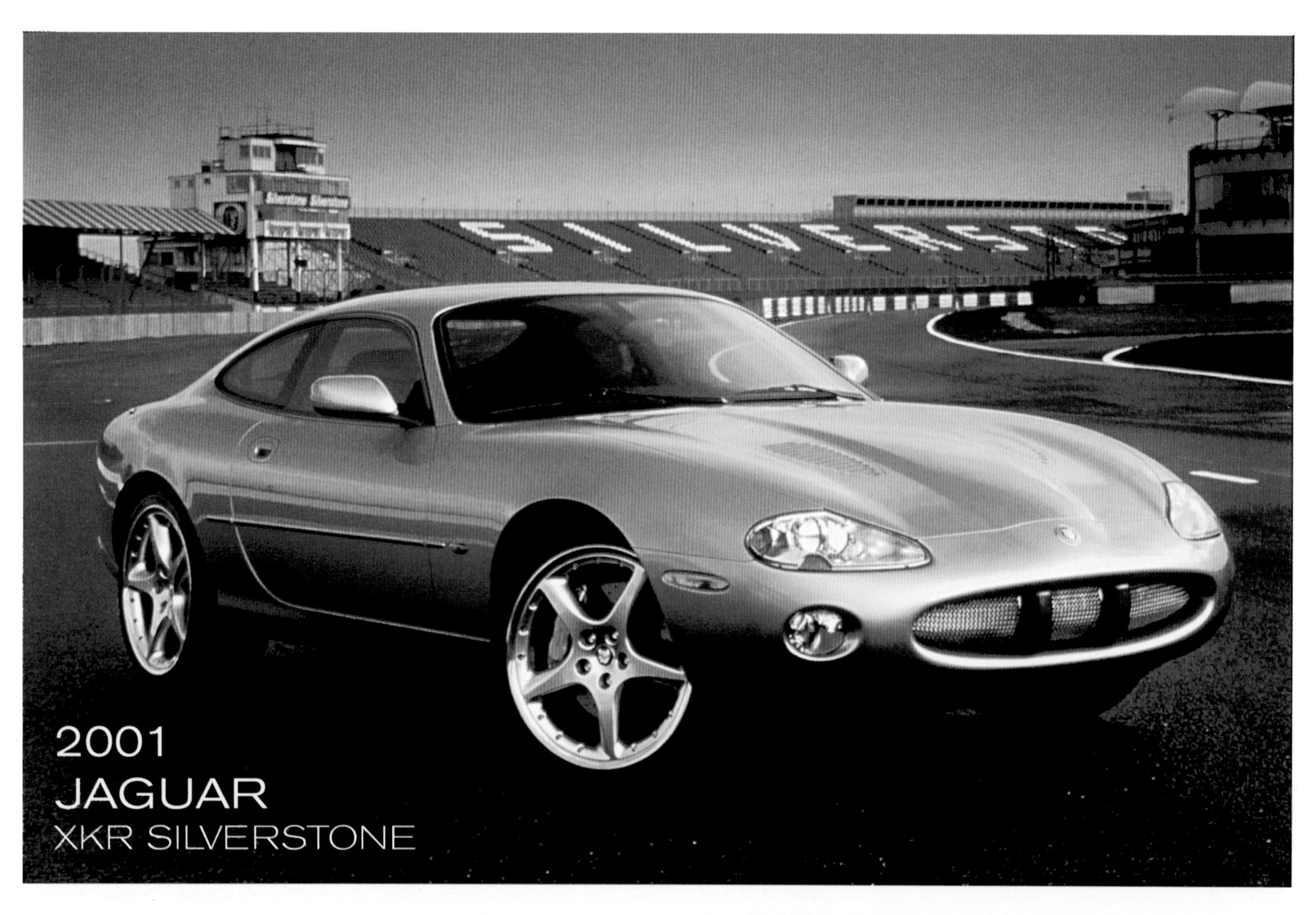


2001
JAGUAR
XKR SILVERSTONE

In 2001, Jaguar brought out a Silverstone edition of its grand touring XKR coupe and convertible. To XKR's equipment the Silverstone added: Platinum Silver paint, "Silverstone" writ on hood emblem and chrome door-sill plates, bird's-eye maple interior planking instead of the usual burled walnut, charcoal leather upholstery with red stitching, larger-diameter all-disc brakes by Italy's Brembo, and 20-inch wheels wearing bigger tires. The coupe chassis was further upgraded with a Performance Handling Pack comprising a slightly larger front antiroll bar, a slightly smaller rear bar, higher-rate springs, and steering with a recalibrated electronic control unit and a rack mounted on firmer bushings.

Named for the famed British airfield circuit where Jaguars raced and won, the Silverstone was an evolution of the XK-Series, which bowed as the 1997 XK8 with Jaguar's then-new 32-valve, 4.0-liter, twincam V-8 and styling with strong overtones of Jaguar's storied E-Type sports cars. The front suspension and part of the floorpan also looked to the past, being held over from the superseded XJS that dated back to 1975.

The XK8 was too heavy and luxurious to be the "sports car" Coventry said it was. But the styling implied otherwise, prompting *AutoWeek's* Pete Lyons to ask, "Is the XK8 really a neo-E? Sorry, no. . . . Then is it just a rebodied XJS? No, it's better, much better. The lively new XK8 can honestly call itself a driver's car—a fine modern GT."

Traces of XJS were in it, but the styling was drop-dead gorgeous; handling was enjoyably agile and the V-8 was a silky, whispering dynamo. But some people, including most journalists, thought the XK8 deserved more than 290 bhp. Coventry added a blown V-8 for 2000 and created the XKR that packed 370 bhp and 387 pound-feet of torque. The

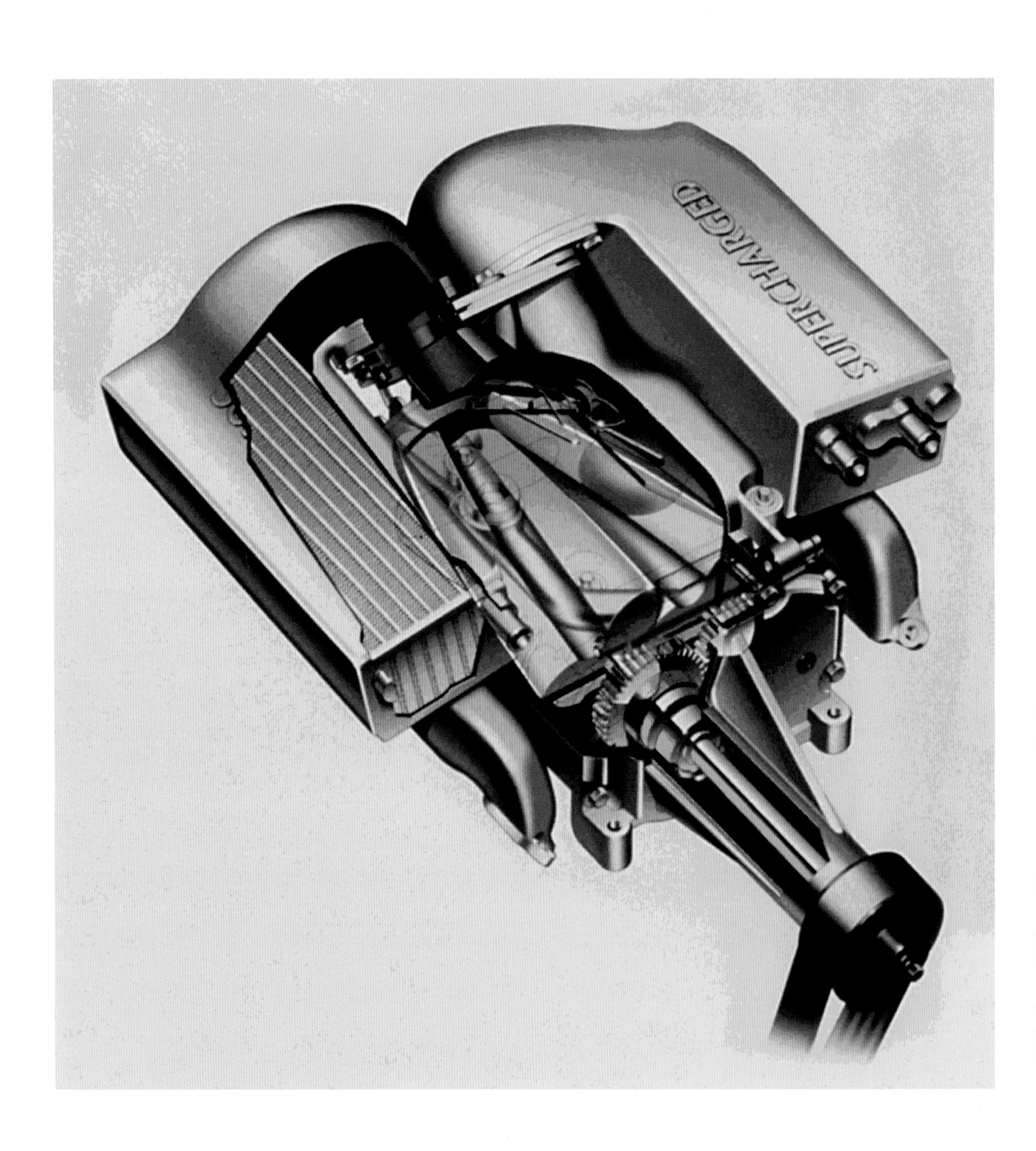
SUPERCHARGED

supercharger made a muted, but discernible, full-throttle moan that recalled the bellow of classic prewar supercharged machines.

Despite its sporty elegance and ineffable "Olde English" Jaguar charm, the XKR proved an exhilarating ride. Most published road tests listed 0-60 mph at about 5.3 seconds regardless of body style. The standing quarter-mile came up in as little as 13.7 seconds at 105 mph. Top speed was electronically limited to 156 mph, but disconnecting the governor might have increased that to 170. *Car and Driver's* test coupe generated 0.88g on the skidpad, not Corvette grippy perhaps, but impressive for a fast, smooth-riding GT weighing nearly two tons.

All these stats naturally applied to Silverstones, which weighed no more than like-equipped XKRs. Even so, the bespoke models had an edge in braking (somewhat shorter) and handling (even more adjustable). After testing one for its December 2000 issue, *Road & Track* reported "the Silverstone feels more surefooted [on the track] than the stock XKR, exhibiting less side-to-side rolling when pushed hard around tight corners. A simple lifting of the throttle will tuck the front end right back on track to the apex. And staying on the throttle a bit longer out of a corner will easily invoke a light but controllable progression of understeer to help position the car."

# 2004-05 CADILLAC XLR

The Cadillac XLR that debuted in 2004 was a suave Mercedes SL fighter based on Chevrolet Corvette bones. Yet the XLR looked nothing like a Corvette or the SL, had a pure Cadillac heart, and was plenty fast. The XLR was basically the for-sale version of the 1999 Evoq concept car, the first public hint that Cadillac's future in the 21st century would not be like Cadillac's past. At the time the Evoq broke cover, General Motors's luxury brand was some two years into a $4 billion extreme makeover dubbed "Art & Science."

The XLR was not a Corvette in a Cadillac suit, though the similarities were undeniable. For starters, both had front-mounted V-8s, rear transaxles, and composite-paneled bodies. Apart from looks and intended audience, it was the powertrain that most separated the GM twosome. The C6 Corvette, of course, boasted Chevy's 6.0-liter LS2 ohv V-8 with a stout 400 bhp. The XLR predictably used Cadillac's own dual-cam 4.6-liter Northstar V-8, only it was a re-engineered "Gen II" version developed for rear- and all-wheel-drive models vs. the original Northstar's strictly front-drive applications. Highlights included con-

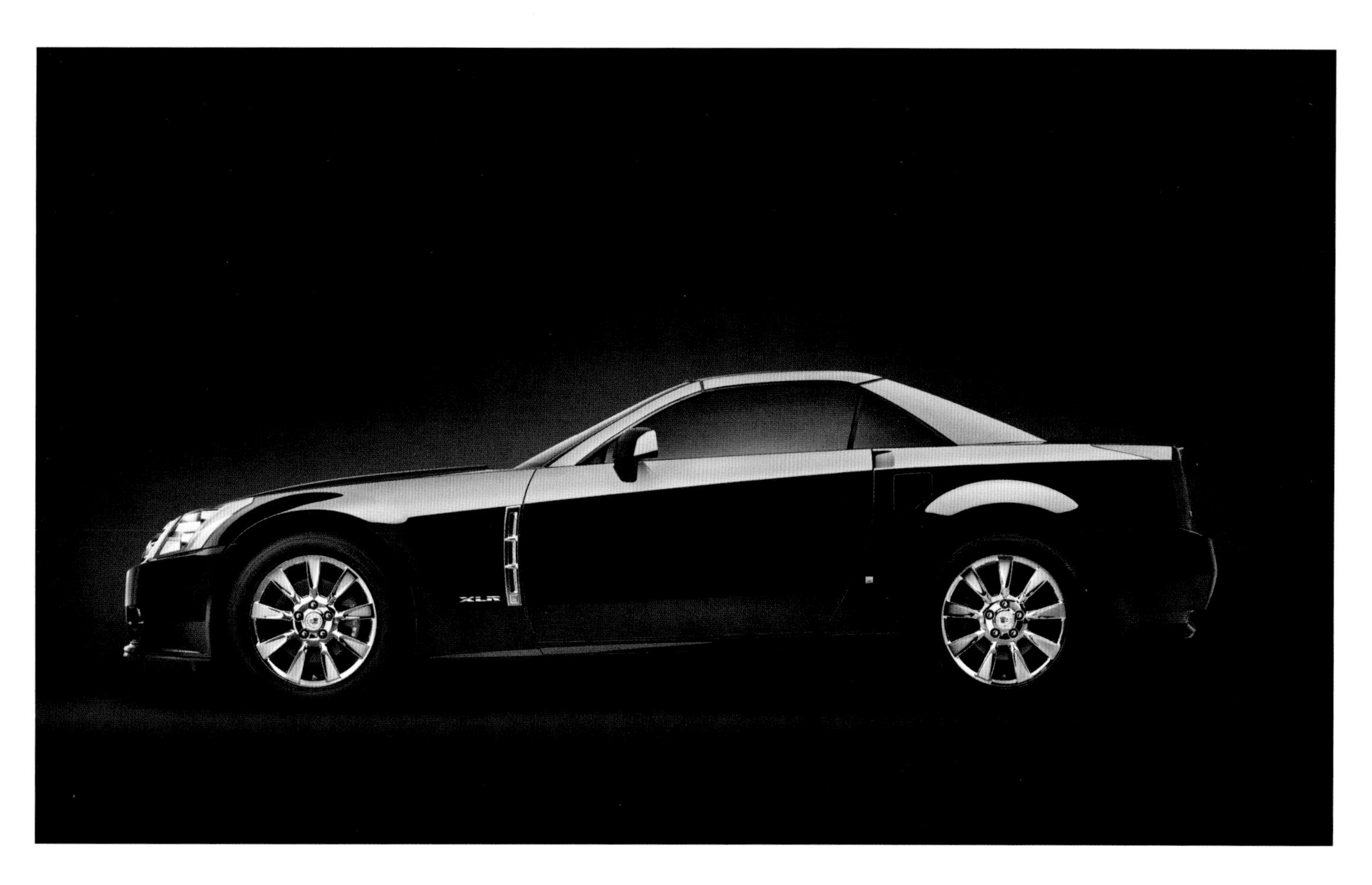

tinuously variable cam timing for both intake and exhaust valves, new cylinder heads with freer-flow ports and higher compression (10.5:1), stiffer block and crank, low-friction polymer-coated pistons, manifolds redesigned for quieter running, and "by-wire" electronic throttle control. The result of all this was 320 bhp at a zingy 6400 rpm, and 310 pound-feet of torque at 4400. Another distinction was the XLR's mandatory five-speed automatic transmission with manual shift gate vs. Corvette's four-speed automatic or six-speed manual. The XLR also replaced the 'Vette's conventional convertible top with a disappearing hardtop and added luxury-class equipment.

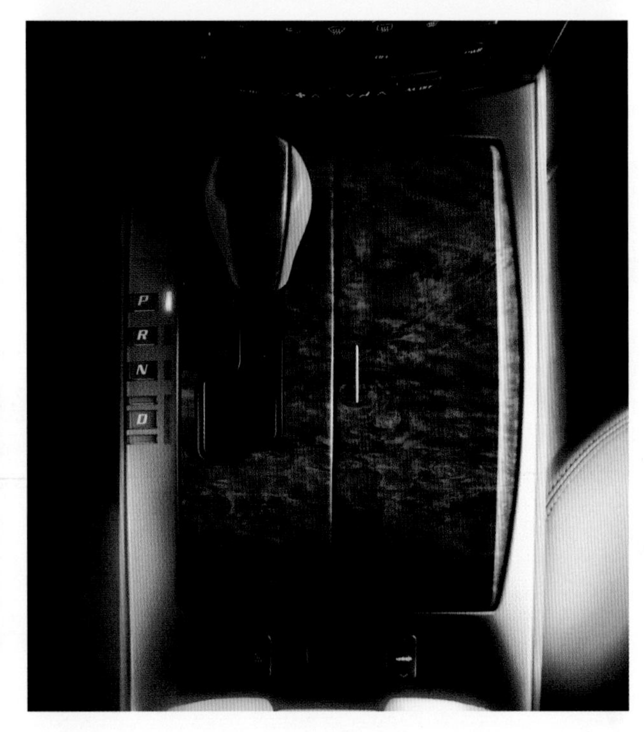


Most every road test at the time judged the XLR one "Xcellent Luxury Roadster." Indeed, the critics praised most everything about it, from the smooth, quiet acceleration—0-60 mph took no more than the factory-claimed 5.9 seconds—to satisfying backroad agility, controlled but comfortable ride, and terrific build quality. Cadillac worked extra hard on that last one, setting up a separate XLR assembly operation that shared only the Bowling Green paint shop and chassis-welding area with Corvette.

There were two bones of contention, though. The power steering still left something to be desired for precision and feedback, and some deemed the run-flat Michelins a bit narrow and soft-sidewalled for best handling. Then again, the XLR hardly wanted for skidpad grip—*Car and Driver* reported 0.83g, *Road & Track* a fine 0.88. A Mercedes SL500 had more cornering power, but was slower in a straight line and cost some $11,000 more.

In all, the XLR was arguably the most impressive new Cadillac since the '67 Eldorado. "[T]here is no question that [it's] a strong entry in the prestigious roadster class," concluded *C/D's* Csaba Csere. "Cadillac's march toward luxury credibility has reached another important milestone."

2005
FORD
THUNDERBIRD

The original 1955-57 Ford Thunderbird was so sleek, so suave, so seductive. Go watch the DVD of *American Graffiti*. Director George Lucas knew the magic of those early 'Birds. Why else would he have chosen one for the mysterious blonde who teases one of the lustful teenage heroes with a fleeting drive-by on cruise night?

The new-millennium 'Bird sought to rekindle that sort of magic for the "happy days" generation, but was also aimed at younger buyers enamored of retro Fifties style. Like the original, it was conceived strictly as a posh V-8 boulevardier two-seater convertible, not a sports car or muscle machine. If you wanted a hot Ford, you got a Mustang GT or Cobra.

The 2002-05 Thunderbird was built on a modified version of a platform shared with Lincoln LS and Jaguar S-Type—Ford owned Jaguar at the time. It also shared a basic rear-drive V-8 powertrain with those sedans. Its 240-cid V-8 developed 252 horsepower in 2002 and got a power boost to 280 for 2003-05. *Car and Driver* found useful performance gains, with 0-60 mph falling from 6.9 to 6.5 seconds (sans hardtop) and the quarter-mile run improved from 15.2 seconds at 94 mph to 15 flat at 95. One could argue that Ford should have gone beyond "relaxed sportiness." Memory, after all, is a tricky thing, and though the new T-Bird was always much faster than its classic forebears, it probably didn't seem so in the minds of some over-50s. However,

if there was such a thing as "heritage feel," this car had plenty. Comfortable and cute, it was just the ticket for people who liked the idea of a sports car without all that pesky sport, much like the Fifties original.

Predictively, the new-century T-Bird boasted amenities never dreamed of in '55: air conditioning with automatic temperature control (dual-zone, yet); keyless-entry power door locks, antitheft system; variable intermittent wipers; and a stereo. Plus, the new 'Bird also had safety features unheard of in the Fifties: airbags, traction control, and antiskid system.

Styling was of the "heritage design" theme. It had things about it you thought you had seen before without being direct copies of things you know that you had seen before. One recreated styling theme was a lift-off hardtop with portholes. The grille, modest hood scoop, and big, round taillights also harked back to the original T-Bird.

The retro 'Birds' Wixom, Michigan, facility was another tie to Thunderbird's past. It was the same plant that opened back in 1958 to build the first four-seat T-Birds and giant unibody Lincolns. Wixom closed in 2007, by which time it was only building Lincoln Town Cars.

Like most "fashion cars," the retro 'Bird fast fell off the sales perch once early adopters got theirs. Two thousand five was Thunderbird's 50th anniversary and final year. There was a 50th Anniversary commemorative fender chevron on every 2005 Thunderbird, plus a 50th Anniversary Edition trim package was available. Just as the original two-seat T-Bird was short lived, so was the retro 'Bird.

# 2008-10 AUDI S5 COUPE

*Road & Track* called the Audi S5 a "hard-charging, road-gripping German tourer of the first order." The S5 went on sale in late 2007 and inaugurated the high-style A5/S5 series that slotted between the compact A4 and midsize A6 groupings. The S5 was the high-performance version of the A5 coupe.

The S5 and A5 were Audi's first sports coupes in 20 years, spiritual successors to the five-cylinder front-drive GT and turbocharged Quattro fastbacks of 1981-91. That first square-rigged Quattro marked the Audi brand's coming of age, winning two drivers' and manufacturers' titles in international rallying and touching off a short-lived craze in AWD road cars that later blossomed anew, particularly among premium brands.

S5 coupes were motivated exclusively by Audi's familiar all-aluminum 4.2-liter (254-cid) twincam V-8 with four variably timed valves per cylinder. In that application it delivered 354 bhp right at

redline, 7000 rpm, and a healthy 325 pound-feet of torque at 3500, aided by a high 11.0:1 compression ratio made possible by direct fuel injection. Transmission choices involved a six-speed manual or ZF's six-speed Tiptronic torque-converter automatic with manual shift gate and steering-wheel paddles.

The S5's styling was crafted by Italian-born Walter de' Silva, who came to Audi after a stint at Alfa Romeo. It was a handsome coupe: chunky, muscular, and well-proportioned, with adept surface development and well-managed reflections. It would have looked even better without B-posts, but that would have required weighty structural reinforcements and the S5 already scaled a hefty 3795 pounds at the curb. (Quattro got part of the blame, as usual.) One nifty touch was the row of LED running lamps curling beneath each headlamp module like Black Bart's mustache.

The S5 was about the same size as a BMW 3-Series two-door and Mercedes-Benz E-Class coupe, standing 182.5 inches long, 73 inches wide, and 53.9 inches high over a 108.3-inch wheelbase. Like most coupes, it was a cozy 2+2, not a practical five-seater. A full-length center console eliminated the rear center position anyway. This relative lack of total passenger space was somewhat offset by Audi's standard-setting cabin execution. That meant the S5 was as lovely inside as it was

outside, with scads of aromatic leather and tasteful dollops of aluminum that could be replaced by optional wood, stainless-steel, or carbon-fiber accents.

For all its heft and luxury, the S5 claimed impressive performance stats. Most road tests reported 0-60 mph in a quick 4.8 seconds or so (with manual), standing quarter-mile runs in about 13 seconds at 105 mph, and more than 0.90g of skidpad grip. What these numbers didn't convey was the smooth sophistication with which the S5 got about its business.

The Audi S5 coupe was one of the most refined, capable, and satisfying 2+2s of its time, not to mention one of the prettiest. *Road & Track* rightly called it "a treat for all the senses."

V8 FSI

## 2009-12
# CADILLAC
### CTS-V

Cadillac's second-generation CTS-V debuted for '09 packing the most-powerful production engine that Cadillac had ever offered: a revised Chevrolet Corvette ZR1 supercharged and intercooled LS9 6.2-liter V-8. The Cadillac variant, dubbed LSA, received unique tuning and components aimed at increasing quietness and refinement, but with little performance sacrifice.

The numbers were jaw-dropping: 556 horsepower at 6100 rpm, and 551 pound-feet of torque at 3800 rpm. For comparison, the 2009 Mercedes-Benz E63 AMG was rated at 507 bhp, the '09 BMW M5 at 500 bhp. If the origi-

ACO 534
CTS

nal CTS-V was Cadillac stomping on its velour-interior/vinyl-top/wire-wheel-cover/V-8-6-4 past, then the second-gen CTS-V was an emphatic grinding of the boot heel.

CTS-V offered a choice of transmissions: a Tremec TR-6060 six-speed manual or General Motors' Hydra-matic 6L90-E six-speed automatic with steering-wheel-mounted buttons for manual shifting.

A handful of performance-themed design alterations differentiated the V from its regular-line CTS siblings. All these changes added a just-right dollop of ominousness to the CTS's already aggressive shape.

Cadillac claimed the CTS-V could accelerate from 0-60 mph in 3.9 seconds and do the quarter-mile in 12 seconds at 118 mph, but magazine road testers generally couldn't replicate those numbers. Most road tests returned 0-60 times in the low-to-mid fours and quarter-miles in the low-to-mid 12s—still world-class. Cadillac engineers aimed for a "bimodal nature" with the CTS-V, so the near-supernatural performance capabilities didn't seriously compromise the luxury and refinement expected of a premium-brand sedan. Outside of a noticeably stiffer ride and a menacing exhaust note, the CTS-V gave up little to

its tamer siblings in mundane stop-and-go driving.

Cadillac was eager to prove the CTS-V's supercar bona fides beyond the expected routine of buff-book road tests and "shootouts." In May 2008, GM Performance Division executive John Heinricy piloted a stock automatic-transmission '09 CTS-V to a blistering 7:59.32 lap of Germany's famed Nürburgring Nordschleife road-race course. Cadillac claimed this was the first sub-eight-minute lap of the 'Ring recorded by a production performance sedan on street tires.

For 2011, a slick coupe and station wagon joined the CTS-V sedan. The coupe and wagon were mechanically identical to the sedan but possessed the obvious benefits and drawbacks of their respective body configurations. Compared to the sedan, both had slightly compromised rear visibility. The wagon came standard with a power liftgate and nearly doubled the sedan's carrying capacity, with 25 cubic feet of cargo space behind the rear seats. With the rear seats folded, that number grew to 53.4 cubic feet.

Regardless of body style, the second-gen CTS-V's "Standard of the World" levels of performance went a long way toward erasing memories of some embarrassing Cadillacs from the then not-too-distant past.

## 2014-19
# BMW
### i8

Beginning with the 2009 introduction of its Vision Efficient Dynamics concept vehicle, BMW began teasing the idea of an exotic sports car with a high-tech, eco-conscious focus. After exploring the concept further (and introducing the i8 name) on a couple subsequent concept vehicles, the company officially committed to a production model. In September 2013, the i8 prototype was unveiled at the International Motor Show in Frankfurt, Germany.

The production i8 launched as a 2014 model, with unorthodox styling that covered an equally unconventional plug-in-hybrid powertrain. And almost all of the concept vehicles' outlandish features—most notably the scissor-wing doors—made the jump from the show floor to the showroom. The starting price was steep ($135,700 in the United States), but not unreasonable compared to similar luxury exotics.

The i8's gas-electric powertrain paired a turbocharged 1.5-liter, 3-cylinder gasoline engine rated at 228 horsepower with a BMW-made synchronous electric motor that was good for 129 hp. Combined maximum output was 357 horsepower and 420 pound-feet of torque. The gasoline engine drove the rear wheels and was mated to a 6-speed automatic transmission; the electric motor sent its power to the car's front wheels via a two-stage automatic transmission. Power for the electric motor was supplied by a liquid-cooled lithium-ion battery pack with a usable capacity

of five kilowatt hours. The powertrain control software allowed the car to be operated on either power source independently or use both of them together, depending on the driving situation; the front-to-rear power split was also variable. BMW quoted the i8's 0-60-mph time at 4.2 seconds, and the EPA's gas-electric mileage estimate was 76 MPGe in combined city/highway driving.

Along with the smaller BMW i3 commuter vehicle, the i8 utilized BMW's LifeDrive architecture concept, which utilized independent "Life" and "Drive" structural modules. In the case of the i8, the Drive module was an aluminum structure that housed the gas and electric motors, the lithium-ion battery pack, chassis and suspension hardware, and the car's crash structure. The Life module consisted of the car's CFRP (carbon-fiber reinforced plastic) passenger cabin. The i8 was 184.6 inches long, 76.5 inches wide, and 51 inches tall, with a curb weight of 3285 pounds.

The interior used a sports-car-typical low seating position. The cabin's appearance followed the layered approach used on the body, along with the imaginative use of contrasting colors. A mix of leathers, cloth accents, painted surfaces, and exposed carbon fiber added to the ambiance. The driver could choose from five driving modes using the Driving Experience Control switch and eDrive button.

*Consumer Guide Automotive* editors tested a 2015 i8, and were impressed by the car's balance of ride comfort and sporty handling, as well as its vivacious acceleration and throttle response. However, the swoopy, low-slung styling and radical doors didn't do any favors for ease of entry and exit or visibility.

BMW announced several updates for the 2019 i8, including the addition of a two-seat roadster variant to be sold alongside the coupe. Other upgrades included an increase of total gas/electric output to 369 horsepower, and an updated lithium-ion battery pack that increased capacity (BMW said net capacity was 9.4 kilowatt hours) and range. There were also some new interior trim choices and exterior colors. Coupe prices started at $147,500, and the new Roadster model had a base price of $163,300.

M BI 8889E

M BI 8889E

# 2018 LEXUS LC 500

Lexus called its LC 500 coupe the luxury brand's flagship. It also said it's a concept car come to life, and in this case it's the LF-LC concept first shown at Detroit's North American International Auto Show in early 2012.

The two ideas certainly went hand in hand. The first thing most people noticed about the LC 500 was its expressive styling. The classic long-hood/short-deck coupe proportions were present, and the racy roofline tapered rearward. Muscular fenders dramatically bulged outward, and bodyside scoops provided an additional sporting touch. Chrome moldings framed the sides of the standard glass or optional carbon-fiber roof panel. Other highlights included elaborately-detailed headlight and taillight assemblies and dramatic pop-out door handles.

The LC 500 was built on Lexus' GA-L ("Global Architecture-Luxury") platform that also underpinned the 2018 Lexus LS sedan. The company claimed this hardware was designed with a low center of gravity as one of its priorities. The body shell was constructed from a combination of high-strength steel, aluminum, and carbon-fiber-reinforced plastic.

The rear-drive LC ran on a 113-inch wheelbase, was 187.4 inches long, and stretched 75.6 inches wide. Curb weight was a substantial 4280 pounds.

The LC 500's extra-long hood hid a 5.0-liter naturally-aspirated 32-valve V-8 engine. It was rated at 471 bhp and 398 pound-feet of torque. Horsepower peaked at a lofty 7100 rpm, with a redline set at 7300 rpm. The engine mated to a 10-speed automatic transmission and was fitted with a standard active exhaust system that allowed the driver to select the level of exhaust sound. Its rambunctious exhaust added a surprising muscle car vibe. Lexus claimed the LC 500 would accelerate from 0 to 60 mph in 4.4 seconds

The standard wheels were 20-inch cast-alloy units, though optional forged wheels in 20- and 21-inch sizes were available too. All tires were run-flats. Chassis highlights included multilink independent suspension front and rear and beefy steel-disc brakes. An active rear steering system was part of the optional performance package. As a 2+2 coupe, the LC 500 was dramatically styled and luxuriously appointed inside. However, that style came at the price of nearly nonexistent rear-seat legroom.

At introduction, the LC 500 was priced from $92,000. Individual extras

included a cold weather pack with a heated steering wheel and windshield deicer, color heads-up display, and a limited-slip rear differential. Option groups were a Touring Package, Sport Package, Sport Package with Carbon Fiber Roof, Performance Package, and Mark Levinson-brand audio.

In a world where the idea of automotive luxury was shifting to SUVs, grand touring coupes still excelled at delivering drama. The Lexus LC 500 could easily cost north of $100,000, but it packed exotic looks, beautifully finished interior, and, of course, ample V-8 power.

# 2018 MCLAREN 720S

Bruce McLaren (1937-1970) was a racing driver from Auckland, New Zealand. He raced for the Cooper Car Company beginning in 1958, and while still with the team started his own outfit, Bruce McLaren Motor Racing, in 1963. The first McLaren racing machine was the M1A sports car of 1964. Formula 1 open wheelers soon followed, along with a series of ever-evolving Can-Am race cars powered by monstrous big-block Chevrolet engines.

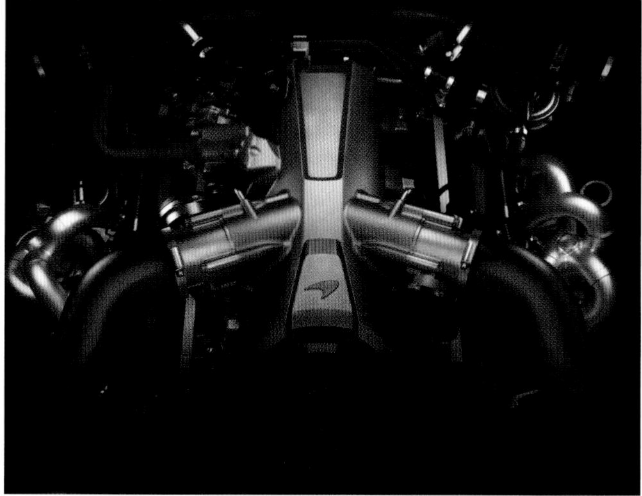

Following Bruce McLaren's death in a Can-Am testing accident, the team continued to race, and went on to achieve victories in the Indianapolis 500 and Formula 1 in the 1970s. Ron Dennis entered the picture in 1980 as the head of an investment group that bought out the company and led the team to even greater success. The first F1 car with a carbon-fiber composite tub, the McLaren MP4/1, debuted in 1981, and the company entered the road-car market with the 1993 McLaren F1. Surrey, England-based McLaren Automotive launched in 2010, and its first road car was the 2011 McLaren 12C.

At the Geneva Motor Show in March 2017, McLaren Automotive introduced its second-generation Super Series car, the 2018 McLaren 720S. It replaced the 650S, and was lighter and faster than its predecessor. Like all other McLaren road cars starting with the F1, the 720S was built on a carbon-fiber tub; this iteration was called the McLaren Monocage II. Bodywork was a combination of carbon fiber and aluminum. The 720S rode a 105-inch wheelbase, was 179 inches long, and had a curb weight of 3128 pounds.

Exterior styling was an evolution of the McLaren look. Highlights included slim roof pillars that enhanced occupant visibility and taut "shrink-wrapped" bodywork. The dihedral doors were double skinned, which allowed them to hide inner ducting that channeled air to the car's radiators and eliminated the need for open side intakes.

The company touted the interior's "perception of space," promising comfortable seating for two along with enough room for an airline carry-on bag behind each seat. Leather trim by Bridge of Weir, machined aluminum switches, and an available Bowers & Wilkins audio system were other high-end accoutrements. The eight-inch Central Infotainment Screen allowed for many of the car's functions to be controlled by touch.

The mid-engine, rear-drive 720S ran McLaren's M480T twin-turbocharged 4.0-liter V-8, which was rated at 710 horsepower and 568 pound-feet of torque. The engine was mated to a 7-speed dual-clutch SSG (seamless-shift gearbox) transmission.

The sophisticated suspension used a double-wishbone setup in both the front and rear, as well as a new generation of McLaren's Proactive Chassis Control (with driver-selectable Comfort, Sport, or Track settings). The brakes paired carbon-ceramic discs with six-piston calipers.

McLaren estimated the 720S was capable of a 0-60-mph time of 2.8 seconds and a quarter-mile sprint of 10.3 seconds. Top speed was a claimed 212 mph. *Car and Driver* was impressed with the car's performance and day-to-day livability, but griped that the touchscreen infotainment system was not up to the standard of "ordinary" cars.

McLaren offered the 720S in three versions; in addition to the standard model, there were Performance and Luxury variants. McLaren also cataloged individual options along with so-called "option packs," and buyers could choose from 20 standard colors. American-market prices started at a cool $288,845.

# 2018
# PORSCHE
## 911 GT2 RS

The GT2 RS was the most powerful and expensive member of Porsche's 911 lineup for 2018, and it was also one of the most ferocious, uncompromising 911s ever built. Racetrack prowess was the GT2 RS's overriding focus, so Porsche prioritized trimming weight wherever possible. Carbon fiber was used extensively; sound insulation was deleted; and a $31,000 Weissach Package added weight-saving gear such as magnesium wheels, carbon anti-roll bars and links, a carbon-fiber roof panel, and even lighter-weight carpet. The air-conditioning and infotainment systems could also be deleted if the buyer desired. The GT2 RS's powerplant was a real monster: a twin-turbocharged 3.8-liter six that made 700 horsepower and was paired solely with a 7-speed automated manual transmission—unlike previous GT2s, a traditional manual transmission wasn't offered. Other high-powered 911s used all-wheel drive to tame power delivery, but the GT2 RS was rear drive only. Performance figures were jaw-dropping: 0 to 60 mph in 2.7 seconds, 0 to 124 mph in 8.3 seconds, and a top speed of 211 mph. Plus, the finely tuned chassis and suspension were up to the task of handling the formidable horsepower. Large carbon-ceramic brake discs, cooled by ducts in the front trunklid and fender extractor vents, provided outstanding stopping power. Active rear-axle steering provided better responsiveness at low speeds and improved stability at high speeds. The GT2 RS proved its potential at Germany's famous Nürburgring-Nordschleife by lapping the 12.8-mile course in 6:40.3 seconds—a record for a road-legal vehicle on that track. Naturally, these superhero capabilities didn't come cheap—the base price of the GT2 RS was $293,200.

D
S GO 5060

# 2019 CHEVROLET CORVETTE ZR1

Chevrolet's Corvette sports car has a long, storied history filled with breathtaking acceleration and handling, groundbreaking styling innovations, and even legendary option codes. One of those codes—ZR1—has come to mean ultimate performance, Corvette style.

The first Corvette ZR1 appeared in 1970, and it was essentially a race-ready option package. Highlights of the package included the solid-lifter LT1 350-cubic-inch small-block V-8, heavy-duty 4-speed transmission, and a tweaked suspension. Since the ZR1 was meant for the track, comfort items like power windows, air conditioning, and a radio weren't available. Chevy produced the ZR1 in very limited numbers from 1970-72.

The ZR-1 name reappeared (this time with a hyphen) for 1990, as the so-called "King of the Hill" Corvette. Technically a $27,016 option for the Corvette coupe, the main attraction was an all-aluminum LT5 DOHC 5.7-liter V-8 good for a substantial 375 horsepower (for comparison, the base L98 5.7 made 245 ponies in 1990). The rear bodywork and rear wheels and tires were wider, and there was a new 6-speed-manual transmission. Styling was touched up for 1991, and for 1993 the LT5's output was bumped up to 405 hp. Production of this generation of the ZR-1 ended after 1995.

The ZR1 was back for 2009, however, based on the aluminum frame developed for the 2006 Corvette Z06. The big news this time around was a supercharged LS9 6.2-liter V-8 rated at an eye-opening 638 horsepower. A 6-speed manual was the only transmission available, and, like the Z06, the ZR1 was offered only as a coupe with a fixed roof panel. The initial base price was $103,300, and production continued through 2013.

2019 CHEVROLET

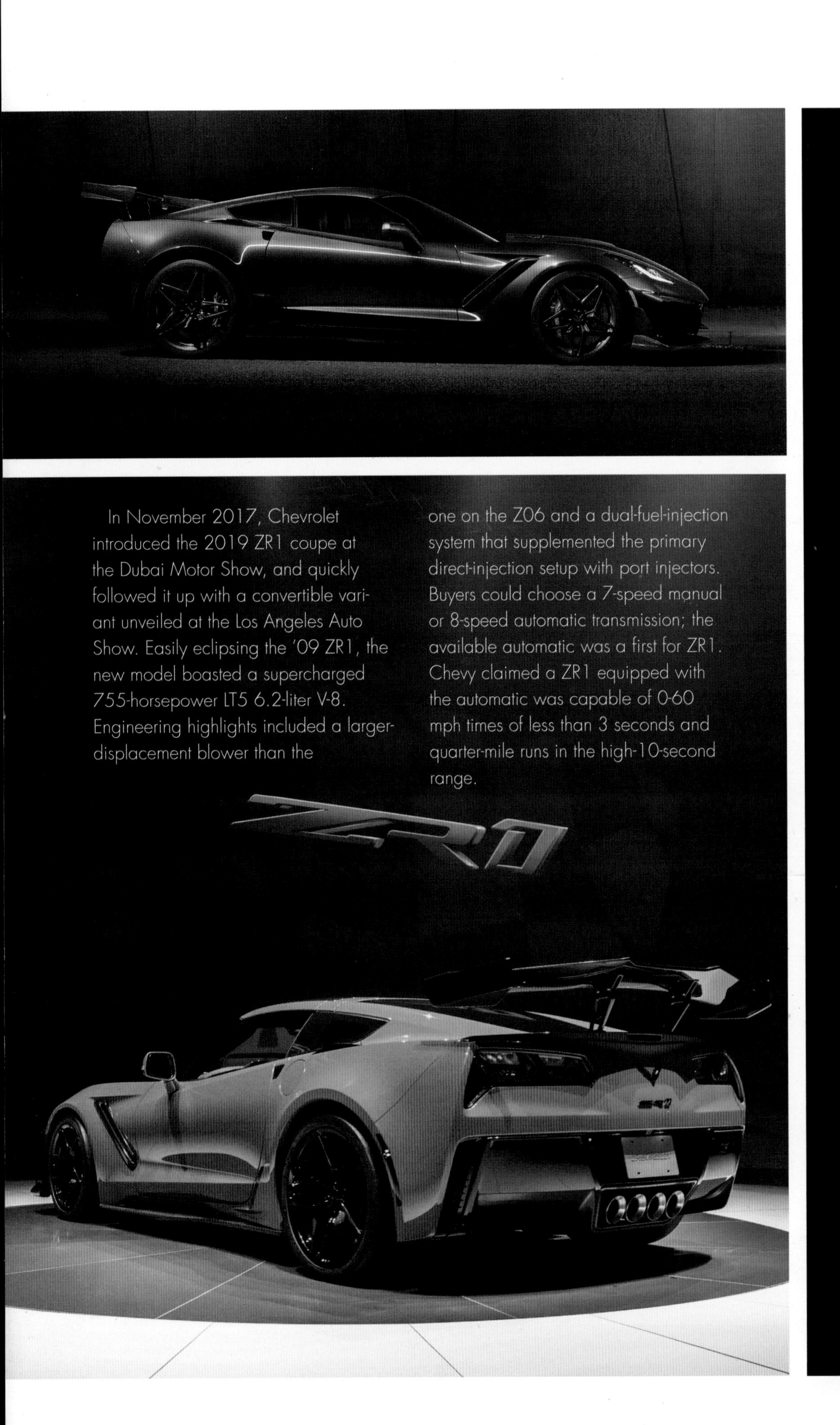

In November 2017, Chevrolet introduced the 2019 ZR1 coupe at the Dubai Motor Show, and quickly followed it up with a convertible variant unveiled at the Los Angeles Auto Show. Easily eclipsing the '09 ZR1, the new model boasted a supercharged 755-horsepower LT5 6.2-liter V-8. Engineering highlights included a larger-displacement blower than the one on the Z06 and a dual-fuel-injection system that supplemented the primary direct-injection setup with port injectors. Buyers could choose a 7-speed manual or 8-speed automatic transmission; the available automatic was a first for ZR1. Chevy claimed a ZR1 equipped with the automatic was capable of 0-60 mph times of less than 3 seconds and quarter-mile runs in the high-10-second range.

Other ZR1 features included an upgraded cooling system that worked with a new front fascia, a specific carbon-fiber hood, and two specially-developed aerodynamic packages. The standard aero setup used a so-called "Low Wing" in back and allowed for the car's highest top speed, claimed to be 212 mph. The optional $2995 ZTK Performance Package added an adjustable "High Wing," front splitter, Michelin Pilot Sport Cup 2 summer performance tires, and specific tuning for the chassis and Magnetic Ride Control. Both rear wings sat on stanchions that were mounted to the car's chassis, similar to the setup on the Corvette C7.R racecar.

There was also a new Sebring Orange Design Package that started with a dazzling Sebring Orange tintcoat paint finish. Orange highlights extended to the brake calipers, accent striping, seat belts, and interior stitch detailing. The 2019 Corvette ZR1 coupe's base price was $119,995, while the convertible started at $123,995.

GT2RS